WORK RELATED INJURY AND ILLNESS LITIGATION HANDBOOK

Timothy Meakin, B.A. LLM
Barrister, Middle Temple

Dr. Peter Ellis, MB.BS (Lond) MRCP (UK)
Barrister, Middle Temple

EMIS Professional Publishing

Work Related Injury and Illness Litigation Handbook
Timothy Meakin and Dr. Peter Ellis

Published by
emis professional publishing ltd
31–33 Stonehills House
Welwyn Garden City
Hertfordshire
AL8 6PU

ISBN 1 85811 315 6

Typeset by Jane Conway

Cover design by Jane Conway
Cover photography by Jon Adams

Printed in Great Britain by Lightning Source

Dedication

Timothy Meakin

To my wife Sarah and our children Lauren, Jack and Evelyn

Peter Ellis

To Rosemary, Natalya and Hamish

PREFACE

Litigation arising out of industrial diseases and injuries presents many challenges, both legal and medical. The prevalence of such claims today, arises in part from historic failures to protect the health and safety of the workforce adequately, particularly in manufacturing industries. This unfortunate legacy means that personal injury lawyers are increasingly obliged to address themulti-faceted complexitites of these claims. Legal issues of foreseeability, causation, limitation and quantum are challenging.

Against this background, the aim of this book is to provide a systematic reference work for this evolving area of the law. In doing so, we have chosen as accessible format in a single volume. The legal chapters set out the fundamental principles, combined with case law and guidance on the practical steps required to bring or defend these types of claims.

In addition, there is a comprehensive summary of the medical issues relevant to industrial diseases and injury claims. This includes the basic principles of anatomy, patho-physiology, diagnosis, treatment and prevention. We have attempted to avoid the jargon which frequently bedevils expert medical reports. However, where this is not possible, glossaries of medical terms, shorthand and acronyms should provide further guidance.

Overall, we have tried to explain the legal and medical issues as clearly and comprehensively as we can, in a non-partisan manner, but also within the confines of a single volume. To the best of our abilities, the law is described as at May 2003.

Timothy Meakin
3 Serjeants' Inn
London

Peter Ellis
Lamb Chambers
Temple

Disclaimer

CONTENTS

TABLE OF CASES

TABLE OF STATUTES AND REGULATIONS

CHAPTER 1
THE LEGAL DUTY OF CARE IN INDUSTRIAL DISEASE LITIGATION

The duty of care at common law: overview

The broad principles underlying a duty of care at common law imposed on an employer are contained in a series of well known previous cases.[1] The issue of the duty of care in the context of liability in negligence was more recently addressed in *Caparo Industries* v *Dickman*.[2] The House of Lords affirmed "the traditional categorisation of distinct and recognisable situations as guides to the existence, the scope and the limits of the duty of care"[3], but there was no expansionist approach to liability. The three tests in law were based on:

(1) sufficient proximity between the parties;[4]

(2) that it is fair, just and reasonable to impose a duty of care; and

(3) that the claimant's injury is reasonably foreseeable.[5]

This test has also now to be considered the context of the Human Rights Act 1998,[6] which is beyond the scope of this book.

1. *Donoghue* v *Stevenson* [1932] AC 562, *Home Office* v *Dorset Yacht Co Limited* [1970] AC 1004, *Anns* v *Merton London Borough Council* [1978] AC 728 HL.
2. [1990] 2 AC 605.
3. At [618].
4. "Proximity" is not defined but it is wider than a spatial relationship and relates more to the directness of the relationship between the parties.
5. *Caparo Industries* v *Dickman* [1990] 2 AC 605 at 617 was applied recently to personal injury cases e.g. *Phelps* v *London Borough of Hillingdon* [2000] 3 WLR 776 (also in this context *X* v *Bedfordshire County Council* [1995] 2 AC 633 HL). See also *Spring* v *Guardian Assurance Plc* [1995] 2 AC 296.
6. See *Osman* v *United Kingdom* [2000] 29 EHRR 23: *Z* v *United Kingdom* [2000] 2 FCR: Article 2 which establishes a right to life and thereby freedom from injury.

The Test of Liability

The nature and extent of the legal duty of care in the context of liability in negligence has been formulated in many previous cases, but two more recent examples are:

(a) In *Page* v *Smith*[7] it was stated that:

> "the test in every case ought to be whether the Defendant can reasonably foresee that his conduct will expose the Plaintiff to the risk of personal injury. If so, then he comes under a duty of care to that Plaintiff. If a working definition of "personal injury" is needed it can be found in Section 38(1) of the Limitation Act 1980. "Personal Injuries" include any disease and any impairment of a person's physical or mental condition."[8]

(b) In *Sutherland* v *Hatton*,[9] the approach in law was put in the following terms:

> "Liability in negligence depends upon three interrelated requirements: the existence of a duty to take care: a failure to take the care which can reasonably be expected in the circumstances and damage suffered as a result of that failure. These elements do not exist in separate compartments; the existence of the duty for example depends upon the type of harm suffered. Foreseeability of what might happen if care were not taken is relevant at each stage of the enquiry."

The Scope of the Duty of Care

Whilst a duty of care may be owed it is nevertheless important to determine the scope of that duty and so further define the limits of foreseeability of the type of harm suffered and its cause. In *South Australia Asset Management Corp.* v *York Montague Ltd*[10] Lord Hoffman used the following example to illustrate the limits of the duty of care:

> "A mountaineer about to undertake a difficult climb is concerned about the fitness of his knee. He goes to the

7. [1995] 2 All ER 736. [1995] 2 WLR 644.
8. Further in law it does not matter if the foreseeable injury was psychiatric or physical.
9. [2002] 2 All ER 1 per Hale L.J..
10. [1996] 3 All ER 365.

doctor who negligently makes a superficial examination and pronounces the knee fit. The climber goes on the expedition which he would not have undertaken if the doctor had told him the true state of his knee. He suffers an injury which is an entirely foreseeable consequence of mountaineering but has nothing to do with the knee. On the Court of Appeal's principle the doctor is responsible for the injury suffered by the mountaineer because it is damage which would not have occurred if he had been given correct information about the condition of his knee.....I think that the Court of Appeal's principle offends against common sense because it makes the doctor responsible for consequences, which though in general terms foreseeable, do not appear to have a sufficient causal connection with the subject matter of the duty.[11]"

A summary of the principles drawn from the analogy cited in *South Australia Asset Management Corp* v *York Montague Ltd* and which have particular relevance in industrial disease litigation is as follows:

(a) The doctor was asked for information about one material consideration as to his safety. There was no good reason why all the foreseeable risks of the expedition should be transferred to him.

(b) A person is under a duty to take reasonable care in the provision of information (or actions undertaken) on which another will rely. He is responsible only for the consequences of the information being wrong (or action is negligent).

(c) A duty of care which imposes on the informant responsibility for losses which would have occurred even if the information which he gave had been correct is not fair and reasonable as between the parties. No tortious duty of care arises in those circumstances.[12]

(d) The criteria for the scope of the duty is limited by reasonable foreseeability of harm. That which is outside the scope of the duty would be rendered too remote to hold the defendant liable.

11. At 214.
12. See also *R* v *Croydon Health Authority* [1998] PIQR Q33: *Darby* v *National Trust* [2001] PIQR 372.

(e) In *Smith* v *Littlewoods Limited*[13] it was held;

> "What the reasonable man is bound to foresee in a case involving damage by independent human agency, just as in cases where such agency plays no part is the probable consequences of his own act or omission, but that in such a case a clear basis will be required on which to assert that the injury or damage is more than a mere possibility."

Foreseeability

A key factor in proving negligence is a reasonable foresight of the kind of damage that in fact happened to the claimant. It is sufficient that a defendant ought to have foreseen the harm to the class of persons.[14] The extent of harm suffered is not a material factor as long as some harm was foreseeable as arising from the breach of duty. The particular factors as to foreseeability of injury, that determine whether the accident falls within the scope of the risk created include:

(a) The precise manner in which the accident happens is not relevant, as long as injury of that kind can be foreseen.[15]

(b) The level of knowledge of the employer of the risk of injury. If an employer did not know, nor could not have reasonably known of the dangers created by the system of work, then the injury was not reasonably foreseeable. This is particularly relevant to claims for asbestosis, mesothelioma, industrial deafness and hand-arm vibration syndrome.[16]

(c) The magnitude of the risk can be sufficient to come within the range of reasonably foreseeable damage unless a reasonable man would ignore it.[17] In *Page* v *Smith*, Lord Ackner stated that the risk of injury was foreseeable:

> "...unless it can be said to be so far fetched or fantastic as to be a mere possibility which would never occur to the mind of a reasonable man."

13. [1987] AC 421 at 461. [1987] 2 WLR 480.
14. As to the test of foreseeability: see *Smith* v *Littlewoods Limited* [1987] AC 421.
15. *Hughes* v *Lord Advocate* [1963] AC 837.
16. This point is expanded upon in the individual chapters on specific industrial diseases.
17. *Overseas Tankship* v *Miller Steamship Co Pty* [1967] 1 AC 617 at 642.

(d) The risk does not have to be more than likely to occur, but has to be a real one as opposed to being a mere possibility of danger.[18] To the extent that physical injury to a person is likely then the damage will invariably be foreseeable.[19]

(e) Psychiatric injury to the victim can be foreseeable and this is the case even if the victim is physically uninjured.[20]

(f) In *Sutherland* v *Hatton*, it was held that the duty on the employer to safeguard an employee from harm from stress at work is based on threshold criteria. The threshold issue is whether this kind of harm to an individual employee was reasonably foreseeable.

The duty of care: the employer/employee relationship at common law

General tortious principles relating to liability are also relevant to the duty of care of the employer in industrial disease litigation. The test is summarised in *Stokes* v *Guest, Keen and Nettleford (Nuts and Bolts) Limited:*[21]

> "the overall test is still the conduct of the reasonable and prudent employer, taking positive thought for the safety of his workers in the light of what he knows or ought to know; where there is a recognised and general practice which has been followed for a substantial period in similar circumstances without mishap, he is entitled to follow it unless in the light of common-sense or newer knowledge it is clearly bad; but where there is developing knowledge he must keep abreast of it and not be too slow to apply it; and where he has in fact greater than average knowledge of the risks he may be thereby obliged to take more than the average or standard precautions."

The Standard of Care: Risk of Injury and Risk Assessment

An employer is not liable in negligence for all the risks that are connected with an employee's work, but only those which should have been acted upon in the appropriate manner.[22]

18. *Smith* v *Littlewoods Ltd* [1987] 2 WLR 480.
19. But see *Bolton* v *Stone* [1951] AC 850. *Mullin* v *Richards* [1998] 1 WLR 1304.
20. *Page* v *Smith* [1995] 2 WLR 644. [1995] 2 All ER 756.
21. [1968] 1 WLR 1776.
22. *Morris* v *West Hartlepool Steam Navigation* [1956] AC 552.

Steps have to be taken to ensure that the system and the place of work are reasonably safe for use by an employee.[23] The standard of care will be judged by that of the reasonably competent employer set in the context of the relevant industry practice. Accordingly the standard will vary with the specialism of the particular trade or industry in question, but employers must attain and maintain a standard of conduct that is reasonable in all the circumstances.[24]

Reasonable Practicability

An employer may contend that the type of business and economic considerations meant it was not reasonably practicable to take the precautions required to prevent an injury to an employee.[25] An employer has to weigh up the risks in relation to the enterprise in question; the greater likelihood of serious injury the greater the corresponding duty on the employer. For example, if there is a risk to life by an undertaking then steps must be taken to prevent it.[26] An employer therefore has to conduct a balancing exercise of the requirements of health and safety and the costs of implementing a safe system of work. Measures should be set in place to prevent foreseeable accidents at work unless the cost of implementing them is "grossly disproportionate" to the risk of injury created.[27] Consequently, in practice effective reliance on the defence of reasonable practicability will be limited.

A summary of legal principles

As the applications of the duty and standard of care of the reasonably competent employer are extensive a summary of the main points on liability is set out below:

(a) The standard of reasonable conduct of the employer is a "moveable feast" and it will depend on the particular facts and conduct in question.

(b) Risk of harm is an important determinant. If the employer did not know, nor ought not reasonably to have known of the risk of injury then it is probably not in breach of its

23. *Wilson's and Clyde Coal Co* v *English* [1938] AC 57: see also Management of Health and Safety at Work Regulations 1999.
24. *General Cleaning Contractors Limited* v *Christmas* [1953] AC 180.
25. See for example: *Morris* v *West Hartlepool Steam Navigation* [1956] AC 522.
26. *Marshall* v *Gotham Co Limited* [1954] AC 360: *Read* v *Lyons* [1947] AC 156.
27. *Coltness Iron Company Limited* v *Sharp* [1938] AC 90.

duty of care. Foreseeability is not made out. But, ignorance of the law is no defence, although in some situations the state of knowledge of risk of injury may only have been known gradually (e.g. hand-arm vibration syndrome).[28]

(c) In contrast where an employer knows of the risk of injury but the risk is one not generally known in the industry then the standard applied will be the personal knowledge of the employer.[29]

(d) If an employer takes reasonable steps to attain and maintain sufficient information relating to the health and safety of the system of work and the place of work then no liability will be made out.[30] The magnitude of the risk is a key determinant of the reasonableness of the conduct of the employer. Whilst the smaller the risk the lower the requirement to act to prevent it is correct, this is qualified by the gravity of the harm done if the risk materialises. The greater the risk of serious injury then the higher the duty imposed on the employer to prevent it, but the risk must be more than a mere possibility.[31]

(e) Health and safety requirements and the reasonable practicability of putting them into effect are relevant.[32] But an omission will rarely be justified where there is a serious risk of personal injury to an employee such that the practice should not be permitted at all.[33]

(f) The costs of precautions and the resources of the employer are relevant. With an employer of limited resources the standard required will be assessed differently to that imposed on a large corporation.[34] Nevertheless a lack of resources will still not relieve an employer of the reasonable standard of competence and so the defence is limited in practice. An employer's resources may be significant in industrial disease and injury litigation and the large employer is expected, for example, to undertake research

28. *Armstrong* v *British Coal Corp* [1997] 8 Med LR 259; [1996] *Times* 6 December.
29. *Stokes* v *Guest, Keen and Nettleford (Bolts and Nuts) Limited* [*supra*].
30. As to the nature of the duty imposed on the employer see *Ball* v *Post Office* [1995] PIQR 5.
31. *Paris* v *Stepney Borough Council* [1951] AC 367.
32. *Latimer* v *AEC Limited* [1953] AC 643.
33. On the practicability of alternatives see: *Nilsson* v *Redditch Borough Council* [1995] PIQR 199.
34. *British Railways Board* v *Herrington* [1972] AC 877.

into injuries and to ensure that all technical equipment is safe to use.[35]

(g) Trade practice of a particular industry is an important factor. But even if trade practice or custom can be proved as supporting the action of the defendant employer, it will not provide a defence to negligence if it is dangerous and likely to cause injury. Its relevance is that it assists the court to assess the industry standard in question.[36]

(h) Whether an employee at known risk of injury ought to be removed or dismissed is ultimately a matter for the employer and not the employee. The fact that the employee knows of the risk of injury but wishes to continue with the work does not necessarily discharge the duty of care. It is a question of the magnitude of risk of injury. The greater the risk of injury the higher the duty on the employer to remove or dismiss an employee for his or her own good.[37]

(i) Health and Safety Executive publications provide a useful indication of the type of risks of injury that an employer ought to have known.[38] Whilst British Standards are not legally binding they provide a reasonable indication of what can be considered best "trade practice".[39] In contrast statutory regulations can be relied on as evidence of "best practice" in a particular trade or industry.

The duty of care: statute

Overview

The duty at common law is mirrored by a series of statutory duties imposed on employers to ensure the safety of employees. The statutory duty imposed will not be subject to the limiting factor of foreseeability and will be absolute unless its terms

35. See *Allen* v *British Rail Engineering Limited* [2001] ICR 942. [2001] PIQR Q10. *Armstrong* v *British Coal Corporation* [*supra*].
36. *General Cleaning Contractors Limited* v *Christmas* [*supra*]; *Morris* v *West Hartlepool Steam Navigation Co. Limited* [1956] AC 552: But see the critical assessment of trade practice in *Thompson* v *Smiths Shiprepairers (North Shields) Limited* [1984] 1 All ER 881.
37. See *Coxhall* v *Goodyear Great Britain Limited* [2002] EWCA Civ 1010: this modifies the principle set out in *Withers* v *Perry Chain Co Limited* [1961] 1 WLR 1314.
38. *Ogden* v *Airedale Health Authority* [1996] 7 Med LR 153.
39. For example, see *Green* v *Building Scene Limited* [1994] PIQR 255.

expressly indicate otherwise.[40] Particular sections of statutes and statutory instruments in the context of industrial disease litigation will be referred to under each chapter, but the statutory regulation that has historically regulated and currently regulates this area of the law includes the following:[41]

(1) Factory and Workshop Act 1901.

(2) Factories Act 1937

(3) Factories Act 1961

(4) Office Shops and Railway Premises Act 1963

(5) The Health and Safety at Work Act 1974.[42]

As to statutory instruments the following represent an essential (but not definitive) list of the relevant regulations:

(1) Asbestos Industry Regulations 1931,1969 (including the Asbestos Prohibitions 1992,1999 and Control of Asbestos at Work Regulations 1987/1998/2002)

(2) Shipbuilding and Ship Repairing Regulations 1960.

(3) Construction (General Provisions) Regulations 1961/ Construction (Working Places) Regulations 1966.

(4) Control of Noise Regulations 1975/1984/1987 (including the Noise at Work Regulations 1989)

(5) Ionising Radiation Regulations 1985/1999/2000

(6) Control of Substances Hazardous to Health Regulations 1988/1994/1999/2002

(7) Construction (Design and Management) Regulations 1994/2000.

(8) Construction (Health, Safety and Welfare) Regulations 1996 The 1992 "six pack" regulations, including the Provision and Use of Work Equipment Regulations 1992/1998, the

40. *Stark* v *Post Office* [2000] ICR 1013. [2000] PIQR 1905.

41. This is not an exhaustive list, and the relevant statutory regulations are considered in the specific chapters which follow.

42. Although by s.47, breach of this statute imposes no civil liability.

Workplace (Health, Safety and Welfare) Regulations 1992, the Health and Safety (Display Screen Equipment) Regulations 1992, the Manual Handling Operations Regulations 1992,[43] and the Personal Protective Equipment Regulations 1992/1999.

(6) The Management of Health and Safety at Work Regulations 1992/1994/1999 (which require a risk assessment in relation to the activity undertaken).

The European Law Context

More recent statutory regulation emanates substantially from European Directives that have been incorporated into UK law and all legislation ought to be purposively construed so as to conform with these obligations, as set out in the directives.[44] The purposive interpretation of EC law has been subsumed within domestic UK law to the extent that where an absolute duty is imposed by EC Directive then this modifies the common law (for example on the issue of reasonable foreseeability). For example it was held in *Stark* v *Post Office*[45] that there was an absolute obligation on the employer under the Provision and Use of Work Equipment Regulations 1992 Reg.6(1) to take measures to maintain the safe and efficient operation of all equipment.[46] The defendant was in breach of those obligations, notwithstanding it could show the injury was not reasonably foreseeable. The relevant Directive, 89/655, set a minimum standard in terms of the duty of employers to ensure the health and safety of employees and the duty was absolute, as set out in the regulations. The practical effect of the implementation of EC Directives is therefore to widen the duty of care imposed on employers under relevant statutory regulation on health and safety law.

Summary

The schedule provides a summary of the main points of this chapter.

43. As to the duty of care in manual handling cases prior to the regulations see *Ball* v *Post Office* [1995] PIQR 5.
44. See *Lister* v *Forth Dry Dock Engineering Co. Limited* [1990] 1 AC 546. [1989] 2 WLR 634.
45. [2000] ICR 1013: see also *King* v *RCO Support Services Limited* [2001] ICR 608.
46. Reg 6 has no qualification on the duty by "reasonable practicability" and it held that the Directive implemented a minimum standard.

<table>
<tr><th>Point</th><th>Comment</th></tr>
<tr><td>• Duty of care in general terms is based on 3 tests in Caparo v Dickman.
• Specific duties in employer/employee relationship are based on a prudent and reasonable employer.</td><td>• Sufficient proximity
• Fair just and reasonable to impose duty
• The claimant's injury must be foreseeable Stokes v Guest et al: provides the standard test of the duty of care. In many cases a duty of care is not an issue as the duty is based on employer/ employee relationship.</td></tr>
<tr><td>Scope of the Duty: Foreseeability
• The duty is limited to the type of harm that could reasonably have been foreseen.
• The precise manner in which the harm happens is not relevant as long as some injury is foreseeable.
• Knowledge of the risk of injury by the employer is a central factor in industrial disease litigation.
• Risk of injury is relevant to the duty of care, including magnitude of risk, severity of injury if it occurs, chance of occurrence. Employers can seek to rely on "reasonable practicability" defence.</td><td>South Australia Asset Management Corp v York Montague Ltd.
Banque Bruxelles v Eagle Star Ins. Co. Ltd
Sutherland v Hatton

• Employers will probably not succeed on a "reasonable practicability" defence where injury is foreseeable based on previous case law.</td></tr>
<tr><td>Statutory Duty
• Imposes additional duty and can modify duty at common law.
• Must be seen in the context of European Directives and minimum standards.
• Foreseeability is not a relevant issue under a statutory duty unless expressly provided for. If the duty is absolute then defendant has to prove compliance.
• The regulations dictate the duty in terms of precautions that should be taken and the employer is under a duty to comply with them.</td><td>Stark v Post Office and consider the EC Directives in setting minimum standards that control the extent of foreseeability of harm.</td></tr>
</table>

CHAPTER 2
CAUSATION IN INDUSTRIAL DISEASE LITIGATION

Overview

A claimant has to prove on the balance of probabilities that the injury suffered was caused by the breach of duty of the defendant, although the conventional categories of the duty of care, foreseeability and causation can overlap to a considerable degree. Causation is in practice invariably determined by foreseeability and remoteness:

> "Three questions, duty, causation and remoteness run continually into one another. It seems to me that they are simply three different ways of looking at one and the same problem. Starting with the proposition that a negligent person should be liable within reason for the consequences of his conduct the extent of his liability is to be found by asking the one question: is the consequence fairly to be regarded as within the risk created by the negligence? If so the negligent person is liable for it but otherwise not."[1]

Causation has also to be seen in a wider context, namely in terms of limiting the responsibility for the harm done by the defendant. In *South Australia Asset Management Corporation* v *York Montague Limited*[2] it was stated in the following terms:

> "Rules which make a wrongdoer liable for all the consequences of his wrongful act are exceptional and need to be justified by some special policy. Normally the law limits liability to those consequences which are attributable to that which made the act wrongful. In the case of liability in negligence for providing inaccurate information this would

1. *Sutherland* v *Hatton* [2002] IRLR 263. [2002] 2 All ER 1.
2. [1996] 3 All ER 365.

mean liability for the consequences of the information being inaccurate."[3]

Proof of causation

What a Claimant has to Prove

Proof of a breach of duty of the defendant is not sufficient, as causation (namely loss or damage flowing from the breach of duty) is also an essential component. The claimant has to prove on the balance of probabilities that the breach of duty of the defendant caused the injury, as opposed to the claimant's work merely being connected with the injury.[4] From this proposition the following points emerge:

(a) Causation in fact has to be proved and this is based on either the act or omission of the defendant. In many cases a series of causes could have been responsible for the claimant's injury and exclusion of the non-causative ones is achieved by applying the "but for" test.[5] If the outcome would not have been affected by the negligence in question, then the case fails the "but for" test and so fails on causation.[6]

(b) In industrial disease litigation, a claimant may prove an exposure to a hazardous substance in the course of the employment in question, but that does not necessarily prove that the hazard has actually caused the injury. If there is no proof of a causal connection (and this means in practice showing the mechanism) then the claim will fail.[7]

---

3. See *Darby* v *National Trust* [2001] at 378.
4. As to the burden of proof: see *Holtby* v *Brigham and Cowan* [2000] 3 All ER 421; *Bonnington Castings Limited* v *Wardlaw* [1956] AC 613.
5. See *Sidaway* v *Board of Governors of the Bethlem Royal Hospital and The Maudesley Hospital* [1985] 2 WLR 480 *Hotson* v *East Berkshire Area Health Authority* [1987] 1 All ER 210 [CA] [1987] 3 WLR 232 as examples of this test. Note also that "injury" includes psychiatric injury: *Page* v *Smith* [1996] A.C. 155.
6. *McWilliams* v *Sir William Arrol and Co Ltd* [1962] 1 WLR 295.
7. *Reay* v *British Nuclear Fuels Plc* [1994] PIQR 171: which related to injury caused by ionising radiation.

Material contribution to injury and apportionment

A Material Contribution to the Injury

In cases of industrial disease, where there can be competing causes, a more sophisticated approach is required to causation and this is reflected in two important cases:

(a) In *Bonnington Castings Limited v Wardlaw*,[8] the employee worked in a dressing shop of a foundry. He contracted pneumoconiosis caused by the inhalation of particles of silica from which he should have been protected. It was held that the machinery at the foundry contributed an amount of silica dust which was more than negligible and so contributed to the disease. The breach of the defendant had made a "material contribution" to the injury, although it could not be proved as wholly causative. But any contribution which was not *de minimis* was "material".[9]

(b) In contrast in *McGhee* v *National Coal Board*[10] the claimant emptied pipe kilns at the defendant's brickworks. He suffered skin irritation allegedly from lack of washing facilities. It was held that where there was an injury with two or more cumulative causes and, one or more of the breaches caused the injury, and one or more did not (so that a proportion could not be ascertained), then the claimant did "not have to prove the impossible". A claimant succeeded where it was shown that the breach of duty was based on a "material increase" in the risk of injury. Such a material increase in risk was to be equated with a material contribution such that causation was made out:

> "...when it was proved on the balance of probabilities that an employer had been negligent and that his negligence had materially increased the risk of his employee contracting an industrial disease he was liable in damages to his employee if he did contract that disease, notwithstanding that he was not responsible for other factors which materially contributed to the disease."[11]

8. [1956] AC 615.
9. See also: *Nicholson* v *Atlas Steel Foundary and Engineering Company* [1957] 1 WLR 613.
10. [1973] 1 WLR 1 HL.
11. CA commenting on Lord Salmon in *McGhee* at para 47.

These two cases showed differing approaches to causation with *Bonnington Castings Limited* v *Wardlaw* requiring proof of a material contribution to the injury by the alleged breach. The approach of *McGhee* v *NCB* was more relaxed and equated a material increase in risk of injury with an actual material contribution to the occurence of the injury, which is clearly logically unsustainable. Nevertheless this approach permitted the claimant to claim full compensation when it was evident that the claimant had not discharged the burden of proof on the medical evidence and could not do so. This was the case even where there were competing non-negligent causes and causation of the injury could not be definitively proved. Effectively, if the defendants wished to deal with respective shares of responsibility under a contribution notice then that was a matter for them. It is the "McGhee principle"[12] that has recently been revived and will significantly affect the practitioner's approach to causation in this area of the law.

A revision of causation

Recent case law has revised the approach to causation and has highlighted more rigorous scrutiny of causation in industrial disease litigation. The approach in summary is as follows:

(a) Separate off conventional causation from cases based on mesothelioma (or those where medical science cannot provide clear evidence on medical causation). These are considered in their own category (see below).

(b) The *Bonnington Castings Limited* v *Wardlaw* approach highlights the requirement, namely that the claimant must prove the negligent system of work made a "material contribution" to the injury. This is the starting point on causation.

(c) Cases have to be categorised as either being "divisible or "indivisible" in terms of their causes. Namely can the cause of the injury be separated out into proportionate percentage allocations of blame.[13]

12. In *Wilsher* v *Essex AHA* [1998] A.C. 1074 it was specifically held that *McGhee* made no new "principle" of law.

13. For example: "divisible" injuries are those that are cumulative such as asbestosis, where the different causes can be identified and assessed as opposed to an "indivisible" injury like mesothelioma which is "all or nothing" in terms of pathology. See *Fairchild* v *Glenhaven* [2002] 1 WLR 1052 [2002] CA ICR 412. See also: *Gregg* v *Scott* [2002] EWCA Civ 1471 *Tahir* v *Haringey H.A.* [1998] Lloyds Rep Med. 104.

(d) For divisible injuries (e.g. asbestosis, which is progressive) the defendant(s) is only liable to the extent of the contribution to the injury caused by the breach of duty and the principles highlighted in *McGhee* v *NCB* cases are not followed.[14]

(e) In contrast, with indivisible injuries the defendant is liable for all the injury and the claimant is entitled to full compensation from that defendant.

(f) As to mesothelioma cases, these are placed in a special category and causation is assessed on new rules. The approach set out by the House of Lords in *Fairchild* v *Glenhaven Funeral Services Limited*[15] is limited to diseases like mesothelioma and the criteria are not of broad application in other areas of causation.

Apportionment of Causation of Injury

The *ratio* of *McGhee* v *NCB* was that there would be no apportionment of liability as to a defendant's actual contribution to the injury. This approach was rejected in *Holtby* v *Brigham and Cowan (Hull) Limited*,[16] where the claimant suffered exposure to asbestos dust in the course of his employment. His employment history was such that he had spent half of it with the defendant, but the remainder with other employers. He developed asbestosis and sued one former defendant employer. The claimant appealed against a 25% reduction of his claim that took account of the exposure to asbestos by other employers. The Court of Appeal rejected the claimant's contentions, that once he proved that the defendant's conduct made a material contribution to his condition, that the burden of proof shifted to the defendant to prove that causation was not entirely made out.[17] In contrast, it held that once it was proved that the defendant's negligence made a material contribution to the claimant's disability, then the defendant was liable only to the extent of its contribution.[18] If quantification was problematic then a "broad brush" was to be applied.[19]

14. See *Holtby* v *Brigham and Cowan (Hull) Limited* [2000] PIQR Q293 sets out the principles: see below.
15. [2002] 1 WLR 1052.
16. Note that no apportionment made in *Bonnington Castings Ltd* v *Wardlaw*.
17. *Bonnington Castings Limited* v *Wardlaw* [*supra*] was again approved.
18. Further the defendant does not have to plead that others were responsible, but it is preferable.
19. As to causation and contributory negligence see: *Stapley* v *Gypsum Mines Limited* [1953] AC 663: *Cork* v *Kirby Maclean Limited* [1952] 2 All ER 402.

Allen v British Engineering Limited: Five Propositions on Causation

The *ratio* set out in *Holtby* was followed in *Allen* v *British Rail Engineering Ltd.*[20] The claimant suffered a vibration white finger injury in the course of his employment. He had worked with vibrating tools over the previous forty years and this included a period with the defendant employer. At first instance a reduction was made of 50% that took account of the apportionment of liability between various employers. On appeal the issues were first, whether the defendant employer was liable solely for the additional injury that occurred as a result of exposure in the index period of employment or for all the injury suffered. Second, whether the claimant would have suffered injury in any event even if he had not been employed with the defendant employer.

Five propositions were laid down by the Court of Appeal and in summary they are:

(1) A claimant has to prove that a defendant employer's negligence made a material contribution to the employee's injury.

(2) In some cases two tortfeasors could have caused the injury by acting in combination over a sustained period of time.

(3) An employer's liability is limited to the extent its negligence added to the claimant's disability.

(4) An apportionment of liability is acceptable, but this is qualified in that a claimant's case should not fail if he cannot establish accurately the apportionment of injury caused between respective employers.

(5) Evidence to prove apportionment must be proportionate to the amount at stake and so the standard of proof of injury in fact has to be kept in perspective.[21]

The approach to causation was reinforced in *Sutherland* v *Hatton*[22] where causation was considered in the context of psychiatric injury. Hale LJ considered the case law of *Holtby* v

20. [2001] PIQR Q101.
21. For a concise and helpful summary of the law on causation see *Sutherland* v *Hatton* [2002] IRLR 263 at 35. [2002] 2 All ER 1.
22. [*Supra*].

Brigham and Cowan Limited where there was an apportionment of the harm caused by each tortfeasor and then drew the distinction with an "indivisible injury":

> "It is different if the harm is truly indivisible: a tortfeasor who has made a material contribution is liable for the whole, although he may be able to seek a contribution from other joint or concurrent tortfeasors who have also contributed to the injury."

Causation in mesothelioma cases

Fairchild v Glenhaven Funeral Services Limited

In terms of causation mesothelioma cases are treated differently to other industrial disease cases since *Fairchild* v *Glenhaven Funeral Services Limited.*[23] The case related to claimants who had all contracted mesothelioma in the course of their employment through exposure to asbestos. All the claimants suffered injury at different times whilst working for different employers and the appeals were all consolidated. In the Court of Appeal the problem on causation that was posed was whether the conventional approach, that could leave an employer 100% liable for the injury where no other insured employer was in existence, was fair. Further, where the medical evidence could not state on balance that a particular exposure materially contributed to the disease.

The Submissions of the Parties in the Court of Appeal

In order to clarify the issues on causation the competing arguments of the parties are set out in summary form.

The claimants contended as follows:

(a) the causal agent (asbestos) had been proved to result in the injury (mesothelioma) and a more precise causal relationship was not required in law, i.e. to show precisely how mesothelioma was caused.

(b) The principles founded in *McGhee* v *NCB* determine whether a claimant has established liability, namely

23. [*Supra*].

whether the tort has materially increased the risk of injury. All in all a common sense approach from the judge was required and the present case was indistinguishable from *McGhee* v *NCB*.

(c) Policy implications meant that if the tort system was to prevent risks to health it should provide compensation where those risks occurred.

The defendants contended that:

(a) Mesothelioma was probably just as likely caused by a single fibre – as by more than one. It was just as likely that different fibres were involved at each of the six or seven stages of genetic change. There was just as much chance of the "guilty fibre" coming from the 10% chance of developing in the disease from background exposure. Nothing could be proved on the balance of probabilities.

(b) *McGhee* v *NCB* should be distinguished in the present case. In *McGhee* v *NCB* there was no doubt that the causative agent was dust which got into the claimant's skin from a day's work. In this case, no-one knew where the fibres came from and there could not be a legitimate inference drawn in favour of the claimant.

(c) It would be unjust to treat this as an indivisible injury and use the "evidential gap" inference to hold one defendant liable for 100%. This was not a "cumulative case".

The Findings of the Court of Appeal

The Court of Appeal held in favour of the defendants. In assessing causation a distinction was to be drawn between indivisible injuries (e.g. mesothelioma) and divisible injuries (e.g. asbestosis). As mesothelioma was an indivisible injury, the claimants could not prove on balance which period of exposure caused or materially contributed to the cause of the mesothelioma. A material increase in the risk of injury through exposure was not sufficient to prove causation. Moreover, the claimants knew it was a matter of chance as to which defendant remained in existence and was worth suing. Effectively, there was a "leap over the evidential gap not only that defies logic but is also susceptible of unjust results."[24] Accordingly, *McGhee* v *NCB* established no new principle of law and in support of that finding reliance was placed on *Wilsher* v *Essex Area Health Authority*.[25]

The Ratio of the Court of Appeal Decision

Whilst the House of Lords overturned the findings of the Court of Appeal there is still a series of important points to be drawn from the judgment that remain of general application:

(a) Where more than one employer has during a certain period exposed the claimant employee to a risk of injury for example to asbestos, the burden of proof remains on the claimant.

(b) In the context of divisible occupational diseases (e.g. asbestosis/pneumoconiosis) a claimant can claim against each or all of the defendant employers in proportion to their contributions. Further, damages will be reduced to reflect the contribution of non-negligently caused injury.

(c) Liability under the Occupiers' Liability Act 1957 would not be made out against an absentee occupier who has delegated work to an independent contractor. The presence of the claimant on the premises (and so its condition) did not cause the injury, but rather the activities carried out on it.

The Decision of the House of Lords

The approach of the House of Lords to the problem of causation in mesothelioma cases was radically different to that of the Court of Appeal. The House of Lords defined the issue thus:

> "The overall object of tort law is to define cases in which the law may justly hold one party liable to compensate another. Are these such cases? A and B owed a duty to C to protect C against a risk of a particular and very serious kind. They failed to perform that duty. As a result the risk eventuated and C suffered the very serious harm against which it was the duty of A and B to protect him. ...If the mechanical application of generally accepted rules leads to such a result there must be room to question the appropriateness of such an approach in such a case."[26]

24. *Per* Brooke LJ.
25. [1988]AC 1074 HL.
26. Per Lord Hoffman.

What was the legal issue to be addressed?

The House of Lords approached the problem in the following manner: a claimant was employed by two employers, but both were under a duty of care to prevent him or her suffering from the effects of inhalation of asbestos dust. Assuming a claimant's mesothelioma was probably caused by inhalation of asbestos dust (as opposed to another cause), but he cannot prove the inhalation of asbestos dust caused the mesothelioma can that claimant still claim against either or both of the employers?

The limitations of the "But For" Test

In seeking to find a solution to the problem the "but for" test was considered relevant but an overly mechanistic approach to the issue of causation was discouraged.[27] There was a need to place the problem of causation in its proper legal context and Lord Hoffman put it thus:

> "The first point to emphasise is that common sense answers to questions of causation will differ according to the purpose for which the question is asked. Questions of causation often arise for the purpose of attributing responsibility to someone, for example, so as to blame him for something which has happened or to make him guilty of an offence or liable in damages. In such cases the answer will depend upon the rule by which responsibility is being attributed."[28]

Therefore a flexible approach is needed as there are usually varying causal requirements. These causal requirements can be difficult to separate from liability as they are intimately interconnected. So first one had to identify the seemingly irreconcilable policy positions of the employer and employee. The House of Lords held as follows:

(i) the claimants' contentions were illogical and susceptible of unjust results, as an employer may be held liable for damage it did not cause;

(ii) the principle that a "material increase in the risk of injury" could be equated with "a material contribution to the

27. Relying on Commonwealth authorities: *Snell* v *Farrell* [1990] 2 SCR 311: Supreme Court of Canada, but also Lord Hoffman in *Environment Agency* v *Empress Car Co, (Abertillery) Limited* [1999] 2 AC 22.

28. *Per* Lord Hoffman.

injury" was logically unsustainable (i.e. the *McGhee* v *NCB* "principle").

(iii) but, there are strong policy grounds in favour of compensating those who have suffered great harm;

(iv) so these shortcomings need to be acknowledged and reappraised, but only in the limited context of this type of industrial disease.[29]

The resolution of the problem

In order to solve the intractable problem of causation in this type of indivisible disease case, the House of Lords held that first, the causal "principle"[30] as set out in *McGhee* v *NCB* must be revised. In order to make sense of causation and produce a just result, one must treat a "material increase" in risk of injury "as if" it was a "material contribution" to the injury.[31] Secondly, causation must be assessed in the context that the potential injustice to the claimant who loses, outweighs any potential injustice to the employer. If the law were otherwise then an employer could escape all liability for mesothelioma. Having identified and modified those points then a logical framework for causation was created. In turn, that framework was based on five specific features.[32]

> "So the question of principle becomes this; in cases which exhibit the five features I have mentioned which rule would be more in accordance with justice and the policy of common law and statute to protect employees against the risk of contracting asbestos related diseases? One which makes an employer in breach of his duty liable for the claimant's injury because he has created a significant risk to his health, despite the fact that the physical cause of the injury may have been created by someone else? Or a rule which means that unless he was subjected to risk by the breach of duty of a single employer he can never have a remedy?"[33]

29. See *Wilsher* v *Essex AHA supra*.
30. Note that in *Wilsher* v *Essex AHA* [1987] *supra* it was emphasised that "*McGhee*" created no new principle of law. *Wilsher* v *Essex AHA* was distinguished in *Fairchild* v *Glenhaven Funeral Services* and in fact said *Wilsher* v *Essex A.H.A.* was wrong on this point.
31. Namely that if one accepts the artificial construct of a causal connection between the two then the issue of causation can applied on that basis.
32. That is, to re-define the causal requirement within new rules.
33. *Per* Lord Hoffman para 63. Reference should also be made to the Speech of Lord Bingham.

The five principles were identified as:

(i) That the duty is intended to protect employees exposed to the risk of a particular disease.

(ii) The duty is intended to create a civil right to compensation for injury connected with the breach.

(iii) The greater the exposure to asbestos the greater the risk of contracting the disease.

(iv) Medical science cannot prove asbestos is more likely than not to have produced cell mutation which caused the disease.

(v) The employee contracted the disease against which he should have been protected.[34]

Summary of the decision of the House of Lords

(1) A new principle relating to causation was created for industrial diseases like mesothelioma, but it is of limited application.[35] Causation is modified for industrial disease claims alone with a resurrection of the *McGhee* "principle". However the principle was not convincingly defined.

(2) The new "principle" relies on the conditions set out by the House of Lords. If the conditions are satisfied, the claimant can recover full compensation against any number of relevant defendants. The relevant conditions are:

- a claimant cannot prove how the injury was caused on account of uncertainty on the medical evidence;

- there had been a breach of duty by the defendant that materially increased the risk of injury (from mesothelioma);

- the breach of duty could cause injury to the claimant;

- proof is required that the defendant's breach was the cause of the injury as opposed to a series of causes;

34. In so finding it was held that the interpretation of Lord Bridge in *Wilsher* v *Essex AHA* [*supra*] on causation and the burden of proof was not binding.

35. See e.g. *Gregg* v *Scott* [2002] EWCA 1471 where Fairchild was held to be not applicable to cases of loss of a chance of a better medical outcome.

- the principle applies where the other possible sources of the claimant's injury include a similar wrongful act or omission of another person, or a non-negligent act of the same defendant.

(3) The "but for" test had limitations and created an unacceptable inconsistency between causation in the case of one employer and the case of two employers. It was unacceptable that D1 should escape liability by reason of a potential (unprovable) claim against D2. The "but for" test was not removed but it does not apply in mesothelioma cases where medical science cannot prove the causal link.

The practical implications on causation

Indivisible and Divisible Injuries

The proper approach to apportioning causation between defendants depends on whether the type of injury is "divisible" or "indivisible" as to its cause. With an indivisible injury (e.g. mesothelioma) there can be no apportionment on causation. It is an "all or nothing situation". In that case the approach as set out in *Fairchild* v *Glenhaven* has to be to adopted. In contrast if the injury is "divisible" (e.g. asbestosis, pneumoconiosis, hearing loss, where there is a correlation between the amount, type and date of exposure and the injury), then an apportionment of causation of injury is applied. In *Holtby* v *Brigham and Cowan Limited* it was stated that:[36]

> "This approach is clearly legitimate in cases of disease or disability where the medical evidence shows that the effect of the defendants' negligence can be treated as cumulative in the sense that the greater the period (or intensity) of exposure, the greater the resultant disease or disability. Asbestosis and pneumoconiosis fall into this category of case. So as Stuart-Smith LJ observed in *Holtby*, does hearing loss, although there the process of just apportionment may be more complex."[37]

36. [*supra*].
37. *Per* Brooke LJ.

Key points

Based on the recent law the practical implications for causation in personal injury litigation in general and industrial disease litigation in particular are as follows:

(a) The key distinction for the approach to causation is whether the injury is "divisible" or "indivisible" and the principles should be applied accordingly.

(b) Practitioners must obtain expert evidence on apportionment of causation as a defendant is only liable for the portion of injury caused. Particular considerations include:

- The precise periods of exposure or relevant employment, identifying the appropriate defendant(s);

- The precise nature of the exposure or work undertaken, demonstrating a logical percentage attribution of causation to the claimant's injury;

- Consideration of what is "proportionate" in terms of proving the claim. Whilst the burden of proof remains on the claimant it is qualified in that the proof required should not force the claimant to prove the impossible.

(c) Tortious and non-tortious causes: this issue has not recently been fully clarified by the Court of Appeal. But on the basis of *Holtby* v *Brigham and Cowan Limited* and *Sutherland* v *Hatton*[38] non-tortious causes (e.g. constitutional conditions, pre-existing conditions etc.) probably need to be taken into account, although there is case law where no such account was taken.[39]

(d) In practice this means that causation must consider pre-existing conditions, that have accelerated or exacerbated a particular condition.

(e) Whilst such factors are taken into account so as to limit causation, there are limits to that approach:

- No account is taken of a vulnerability or pre-existing condition if that would be inappropriate on the facts of

38. [*supra*].

39. *Vernon* v *Bosley [No.2]* [1997] 1 All ER 577 where in a claim for psychiatric injury no account was taken of symptoms arising from a grief/bereavement reaction.

a case (e.g. HAVS, where the use of vibrating machinery has no causal relation to an underlying susceptibility or pre-existing condition);[40]

- The "susceptibility to injury" argument must be seen in terms of the breach of duty. For instance, in HAVS cases, a certain level of exposure may overfill the "reservoir of tolerance" of the claimant causing the presentation of symptoms. However, exposure below that point (and terminated before the reservoir becomes full) may not have resulted in symptoms, as the "the reservoir of tolerance" would not have overflowed;

- In contrast, if the "reservoir of tolerance" of the claimant was not full but the symptoms still presented then causation is not made out fully or in some cases not at all;[41]

- An allegation that an employee would have run the risk in any event and so would have suffered the same injury is not a sustainable contention.[42]

Mesothelioma cases

With an "indivisible injury" like mesothelioma, causation is made out based on the "McGhee principle", namely proof of a material increase in the risk of injury. But even then the further conditions set down in *Fairchild* v *Glenhaven Funeral Services* must be satisfied.[43] If the claimant proves causation then s/he claim full compensation against one or more defendants as s/he does not to have to "prove the impossible" on causation.

40. Effectively anyone who uses this type of machinery for long enough will suffer a HAVS.
41. *Hall* v *British Gas Plc* [1998] Unreported.
42. *Hall* v *British Gas Plc* [*supra*].
43. Note that the principles set out in *Fairchild* are of limited application and most industrial disease litigation will be subject to the principles set out in the *Allen* and *Holtby* cases.

Psychiatric injury cases

Apportionment of Causation

The guidance set out in *Holtby* v *Brigham and Cowan Limited* is probably of wider application and applies in psychiatric injury claims (for example, claims based on stress at work).

In *Coxon* v *Flintshire County Council,*[44] the Court of Appeal upheld the approach and assessment of the trial judge.[45] That approach specifically considered the *Holtby* case. Scott-Baker J held:

> "While I am able to make findings on the issue of causation, there are echoes of the words of Mustill J in *Thompson* in that there is an impossibility of making a precise apportionment between what the Defendant's negligence has caused and what has been caused by other factors. Inevitably, I have taken a broad view and done my best to reach a fair conclusion on the whole of the evidence. It is very much a matter of feel."

The issue was further considered in *Sutherland* v *Hatton*[46] in a stress at work claim. Lady Justice Hale stated:

> "Many stress related cases are likely to have a complex aetiology with several different causes. In principle a wrong-doer should pay only for that proportion of the harm suffered for which he, by his wrong-doing, is responsible."[47]

She then continued by stating:

> "Hence if it is established that the constellation of symptoms suffered by the claimant stems from a number of different extrinsic causes then in our view a sensible attempt should be made to apportion liability accordingly."

Based on those principles it was held that:

(1) Where a tortfeasor's breach of duty has exacerbated a pre-existing disorder or accelerated the effect of a pre-existing

44. [2001] PIQR Q9.
45. This related to severe psychiatric injury caused by sustained sexual abuse whilst in care: General damages PLSA: £35,000.
46. [*Supra*].
47. See also *Rahman* v *Arearose and University College London NHS Trust* (2000) 3 WLR 1184.

vulnerability the award of general damages has to reflect the acceleration or exacerbation. But this assumes that the symptoms caused would have resulted in the injury suffered.[48]

(2) The assessment of financial loss must take account of contingencies. For example, where the claimant would have succumbed to a psychiatric disorder in any event, then this means a reduction of the multiplier.

Summary of this chapter

The schedule contains a summary of the main points of this chapter.

48. For example, if the pre-existing symptoms would have resulted merely in low mood that did not affect the ability to work then the pre-existing condition is not causally relevant to loss of earnings, although it may reduce the general damages.

Issue	**Comment**
The claimant has to prove a causal connection between the breach of duty and the injury. The burden of proof is always on the claimant.	Case law: *Fairchild* v *Glenhaven* *Holtby* v *Brigham and Cowan* *Allen* v *BREL* *Sutherland* v *Hatton* A helpful case on the principles is *Athey* v *Leonati* [1999] Lloyds Rep Med 458
The "but for" test applies in most areas of factual causation but it does have limitations: e.g. mesothelioma cases.	*Fairchild* v *Glenhaven* indicated a refinement of the test in mesothelioma cases.
Causation needs to be seen in terms of "divisible" and "indivisible" injuries as the approach will be different in each case.	This is set out in *Sutherland* v *Hatton* and *Holtby* v *Brigham and Cowan.* Reference also should be made to *Allen* v *BREL* as to the five propositions on causation.
In progressive injuries (e.g. asbestosis) the *McGhee* principle of "material increase in risk satisfies causation" is not followed. Claimant must prove the negligence made a "material contribution" to the injury.	The *McGhee* principle is applied in the limited circumstances of a mesothelioma case.
If a material contribution is proved then the court will then go on to assess the contribution of the defendant. Contributions between tortfeasors will be assessed.	The court will assess the contribution and this is based on medical evidence. Probably applies to non-tortious causes so the claimant's damages are reduced by the contribution of the same to the injury. (This includes exacerbation and acceleration cases.)
In indivisible cases the House of Lords has overruled the Court of Appeal in *Fairchild* v *Glenhaven*. It has set out conditions for claiming in a mesothelioma type case where medical evidence on causation cannot answer the legal questions. It is limited in scope.	This is a new approach and reference should be made to the speeches of Lords Hoffman and Bingham. There may be scope for a wider application of the principle in future.

CHAPTER 3
PROVISIONAL DAMAGES

Overview

A provisional damages award enables a claimant initially to claim damages at trial for injury on the assumption that there will not be a serious deterioration in the condition or disease in the future. But if there is a serious deterioration then a further application can be made for additional compensation. The power to award provisional damages is contained in s.32 Supreme Court Act 1981 and s.51 of the County Courts Act 1984. The procedure for provisional damages, which is a discretionary remedy, is contained within CPR Part 41. Part 41.2(1)(c) defines an award of provisional damages as meaning:

> "an award of damages for personal injuries under which:
>
> (i) damages are assessed on the assumption referred to in SCA s32 or CCA s.51 that the injured person will not develop the disease or suffer the deterioration; and
>
> (ii) the injured person is entitled to apply for further damages at a future date if he develops the disease or suffers the deterioration."

The legal principles

The Criteria For The Award

Provisional damages are a central feature of some areas of industrial disease litigation, specifically asbestosis and mesothelioma cases, where the prognosis is often uncertain. There is a real risk that a claimant's medical condition could deteriorate in the future. If an order is made there is usually only one further application for damages but it can be more than one application. This period can be extended by order of the court. In *Fairchild* v *Glenhaven Funeral Services Limited*[1] it was stated that

> "the court's task is to decide whether there is a chance (proved or admitted) that at some time in the future the claimant will as a result of the exposure to asbestos (while employed by the defendants) develop the identified disease or suffer the identified deterioration."

The award can only be made if relevant conditions are satisfied:[2]

(a) the burden of proof is on the claimant;

(b) the claim for provisional damages has to be pleaded; and

(c) the court is satisfied that SCA s32A or CCA s.51 applies; and

(d) the claim for provisional damages is particularised, including the nature of the disease or condition contended for and the chance of a future serious deterioration in that condition or disease.[3]

If an order is made, then the type of disease or condition has to be specified within the order. Any "deterioration in the condition or disease" has to be causally connected to the original tort of the defendant. In addition, the period which the order will cover has to be defined.[4] In practice the medical evidence has to cover:

(a) the nature of the disease or deterioration of the condition in question;

(b) the chances of a serious deterioration or disease in percentage terms; and

(c) the medical parameters of any such deterioration or disease.

---

1. [2002] 3 WLR 89.
2. *Fairchild* v *Glenhaven Funeral Services Limted* [2002] 3 WLR 89.
3. Medical expert evidence needs to address this point.
4. *Fairchild* v *Glenhaven Funeral Services* [*supra*] considered the issue of provisional damage in the context of an industrial disease claim.

The Conditions: "the Chance of a Serious Deterioration"

In industrial disease claims "the chance" of a serious deterioration in the condition or disease does not have to be high (i.e. more than *de minimis*) and in many cases of mesothelioma and asbestosis courts have made awards of provisional damages. But a second material consideration is whether the condition can in fact be categorised as a "serious deterioration (or disease process)". A distinction is drawn between a "serious deterioration" and a "mere progression" which would not warrant an award. If the deterioration in the condition is not "serious" then likewise, the award is inappropriate.[5] Whether any deterioration is "serious" is ultimately a matter of fact in the circumstances of the case, but a wider interpretation of "serious deterioration" has been advanced.[6]

Death and the Damages Act 1996

The subsequent death of a claimant, who has made a claim, is now covered by s.3 of the Damages Act 1996 (which is retrospective). It states that provisional damages do not preclude a claim under the Fatal Accident Act 1976. Section 3(3) states:

> "(3) Such part (if any) of—
>
> (a) the provisional damages; and
>
> (b) any further damages awarded to the person in question before his death,
>
> as was intended to compensate him for pecuniary loss in a period which in the event falls after his death shall be taken into account in assessing the amount of any loss of support suffered by the person or persons for whose benefit the action under the Fatal Accidents Act 1976 is brought.

5. *Patterson* v *Ministry of Defence* [1986] Simon Brown J: unreported: a progressive condition which would not usually merit an award includes osteoarthritis or similar disability.
6. *Wilson* v *Ministry of Defence* [1991] 1 All ER 636: As to where the claimant is awarded future loss of earnings but if there was a serious deterioration it would mean death and so lead to overcompensation: see *Molinari* v *Ministry of Defence* [1994] PIQR Q33. *Curi* v *Colina* (1998) 14.10.98 held that a "chance" was to be equated with a "possibility but not more than a possibility"; namely "more than fanciful but less than probable."

(4) No award of further damages made in respect of that person after his death shall include any amount for loss of income in respect of any period after his death."

The practical approach to provisional damages

Provisional damages are invariably to a claimant's advantage and a defendant will want to avoid such an order, if possible. Consequently both defendant and claimant practitioners have to ensure that the proper issues are addressed to support their respective positions. To that end the following points are relevant:

(1) The medical evidence has to assess the issue of the "chance" of a "serious deterioration" in the future medical condition or occurence of the disease. The particular requirements of CPR Part 41 need to be emphasised to the respective medical experts and covered in their reports.

(2) Defendant practitioners have to concentrate on whether the medical evidence can show the chance to be *"de minimis"* and so not warranting an award. Further, whether any future deterioration in the condition will not be "serious", such as "mere progression", within the criteria set out in CPR Part 41.

(3) Defendant practitioners need also to consider the grounds on which a final award can be made and the evidence in support of that contention. This relates for example, to whether the medical trigger and/or the period is too speculative and/or whether a provisional order places undue financial hardship on the defendant.[7]

(4) Claimant practitioners need to ensure that the percentage chances of a serious deterioration in a claimant's condition and its nature and extent are comprehensively set out in the medical reports. Further, the expert has clearly to identify the triggers and parameters of the claim on the medical evidence.

(5) Consideration by both claimant and defendant practitioners has to be given to causation in the context of industrial disease litigation. Specifically, the claimant must prove that

7. Note: Part 36 payments can still be made in the context of a provisional damages award.

any deterioration is consequent on the original tort, as opposed to other non-tortious causes.

Case Law

Set out below is a series of recent cases that gives some indication of the approach of the courts to orders for provisional damages in the context of industrial disease litigation.

Case	**Injuries Relating to Provisional Damages Award**	**Was an Order made? and the Terms?**
Little v *VSEL Birkenhead Limited* [1998] CLY 1627.	Exposure to asbestos dust at age 60. Mild asbestosis.	Order made in the event of mesothelioma or lung cancer attributable to the asbestosis or diffuse pleural thickening.
Ward v *Newhalls Insulation Co Limited* [1998] CLY 1623.	Claimant, 46 asbestos exposure. Pleural thickening and restrictive lung disease. Probability of deterioration. 5% chance of lung cancer and 5–10% chance of mesothelioma.	Order made in relation to combination of cancer and mesothelioma risks.
Godfrey v *Gnitrow Limited* [2000] 1 QR 8.	Claimant 57, asbestosis diagnosed and likely to progress. With serious deterioration there was life expectancy reduction of 3–5 years. Lung cancer chance of 3–5% and mesothelioma 10%. Pleural thickening 2%.	Order made in relation to combination of cancer and mesthelioma risks.
Holmes-More v *Tenon Limited* [2001] 3 QR 18.	Claimant 49, diagnosed with mesothelioma. 5% chance of lung cancer and 30–50% chance of pleural thickening.	Order made in relation to pleural thickening and death caused by mesothelioma, or related cancers.

Case	Injuries Relating to Provisional Damages Award	Was an Order made? and the terms?
Fashade v *N. Middlesex Hospital NHS Trust* [2001] *Current Law*: May.	Claimant, 52, suffered bronchial asthma and bronchitis. Respiratory disability at 50–60% at retirement and 7.5% chance of increase to 80% in her mid seventies.	No order made: risk was not "clear and severable" as opposed to a "continuing deterioration."
Molinari v *Ministry of Defence* [1994] PIQR QR 33.	Fitter claimant 39, contracted leukaemia. In remission at date of trial. After 5 years from bone transplant the chance of leukaemia was 1%.	Order made for provisional damages in event of leukaemia.
Bahardo v *Mayer Newman & Co Ltd* [2002] 3 Q.R. 5.	Male, 61 was exposed to asbestos dust. Diagnosed as having asymptomatic pleural plaques and diffuse pleural thickening. Overall respiratory disability was 30 to 40%, but 25% was referable to chronic obstructive airways disease caused by smoking. 20% risk of further pleural thickening causing increased disablement, a 1–2% risk of asbestosis, a 15 to 20% risk of suffering lung cancer. A 8–15% per cent risk of mesothelioma.	Order made in relation to the general damages award was discounted by one-quarter from £27,500 which would have been the award on a non-provisional basis. General Damages: (provisional award) £20,625.

Summary of the Chapter

Set out below is a summary of the main points on the law on provisional damages in addition to practical guidance in relation to making an application.

Issue	Comment
Provisional damages: damages assessed at trial on the basis that there will be no "serious deterioration" in the claimant's condition or disease process. If so, then further damages can be awarded.	SCA s.32A. CCA s.51. Part 41 CPR. The burden of proof is on the claimant. The claim must be pleaded.
Definition and assessment is required of *"Chance" and "serious deterioration"* and these determine whether the award is made. There can usually only be one further award. The conditions by which provisional damages are triggered have to be clearly defined.	"Chance" is not strictly measured but should be more than *de minimis*. *Curi* v *Colina* "Serious Deterioration" is defined in terms of a deterioration that cannot be assessed at trial. If the condition is "progressive" then a medical opinion can probably be given on prognosis and provisional damages are not appropriate. *Wilson* v *Ministry of Defence* is the benchmark case.
Medical evidence is required to define the terms and the triggers for provisional damages.	Conditions, chances and triggers have to be clearly defined by the medical experts in their report.

CHAPTER 4
LIMITATION IN INDUSTRIAL DISEASE LITIGATION: SECTIONS 11 AND 14 LIMITATION ACT 1980

Overview

Limitation issues are relevant to industrial disease litigation in the United Kingdom. Symptoms from more serious industrial disease may lie dormant for many years, or be masked by other non-tortious conditions, thereby preventing a claimant having knowledge of the injury within the conventional time periods.[1] An additional complicating factor is that the Limitation Act 1980 may not apply to the period in which the cause of action accrued and reliance is placed on previous limitation statutes (for example Limitation Act 1939). Consequently practitioners must swiftly identify and deal with limitation issues. Any culpable delay in the litigation may compromise the chances of a successful claim or defence. In this chapter, the general principles are set out in addition to those cases that are relevant to industrial disease and occupational illness litigation.

The Limitation Act 1980

Sections 11 and 12

Section 11(4) of the Limitation Act 1980 ("the Act") states there is a limitation period of three years which starts from

(a) the accrual of the cause of action, or

---

1. As to foreign torts see English Foreign Limitation Periods Act 1984 s.1. As to *forum conveniens* and cases with a foreign element see *Schalk Lubbe* v *Cape PLC* [2000] 1 WLR 154. *Durham* v *T and N PLC* [1996] Lawtel 2.5.96.

(b) the date of knowledge of injury, whichever is the later.[2] Hence much depends on when the claimant actually first had knowledge of the injury in determining when the limitation period commenced and expired. Once the limitation period commences it cannot be stopped. Even if a claim is admitted to be statute barred (thereby necessitating an application under s.33 of the Act) the date of knowledge of injury is still relevant to fix the period of delay since the expiry of the limitation period.

As to fatal accidents,[3] by section 12 of the Act the limitation period is three years from date of death, or three years from the date of knowledge of the person for whose benefit the claim is brought, whichever is the later. If the claim was statute barred before the deceased's death then the provisions under the Act do not alter that position in law. The claim is already statute barred and an application has to be made under s.33 of the Act to disapply the limitation period.[4] In contrast, if the limitation period had not expired at the date of death of the deceased then the personal representatives have three years to commence the action from the date of death.[5] But the provisions relating to knowledge under the Limitation Act 1980 will then apply to the personal representatives.

The Claimant's knowledge of injury under the Act

A claimant has to issue proceedings within three years of the date of

(a) the accrual of the cause of action; or

(b) the date of knowledge of injury, whichever is later.[6] A claimant's "knowledge" of a "significant injury" is critical in assessing when the limitation period commenced and

2. "Personal injury" is defined by section 38(1).
3. Including claims under the Law Reform (Miscellaneous Provisions) Act 1934 by the personal representatives on behalf of the estate.
4. A claim for an indemnity from insurers under the Third Party (Rights Against Insurers) Act 1930 is not a personal injury damages claim. The limitation period is 6 years from the date of judgment: *Burns* v *Shuttleworth Limited* [1999] 2 All ER 27.
5. Section 11(6) 11(7).
6. *Nash* v *Eli Lilley and Co* [1993] 1 WLR 782.

expired. "Knowledge" of injury is further defined in section 14(1) of the Act:

> "In sections 11 and 12 of this Act references to a person's knowledge are references to the date on which he first had knowledge of the following facts:
>
> (a) that the injury in question was significant; and
>
> (b) that the injury was attributable in whole or in part to the act or omission which is alleged to constitute negligence, nuisance or breach of duty; and
>
> (c) the identity of the defendant;
>
> (d) if it is alleged that the act or omission was that of a person other than the defendant, the identity of that person and the additional facts supporting the bringing of an action against the defendant;
>
> and knowledge that any acts or omissions did or did not, as a matter of law, involve negligence, nuisance or breach of duty is irrelevant."

Definitions

In defining the date at which a claimant has "knowledge of a significant injury", such as to trigger the limitation period, the following points are relevant:

(a) "*Knowledge*":[7] means "*to know with sufficient confidence to justify embarking on the preliminaries to the issue of the writ.*"[8] This is a matter of fact for the court to determine but the test is subjective. Once the limitation clock starts ticking it cannot be stopped.

(b) "*Significant*":[9] the injury must be "significant", which means:

> "..if the person whose date of knowledge is in question would reasonably have considered it sufficiently serious to justify his instituting proceedings for damages

7. *Nash* v *Eli Lilley and Co* [*supra*]. See also: *Various Claimants* v *Bryn Alyn Community (Holdings) Ltd* [2003] EWCA Civ 85.
8. *Halford* v *Brookes* [1991] 1 WLR 428.
9. "The injury in question" is defined and considered in *McManus* v *Mannings Marine Limited* [2001] EWCA Civ 1668.

against a defendant who did not dispute liability and was able to satisfy a judgment."[10]

(c) Further, "significant" was *directed solely at the quantum* of the injury and not the claimant's evaluation of its cause, nature or unusualness.[11]

(d) "*Attributable*" is defined as meaning "capable of being attributed to" in the sense of being a real possibility.[12] Further, the claimant has to know the connection between the act or omission of the defendant and the injury (i.e. making a connection between the two). He does not have to know that the act or omission relied upon was negligent. In many cases a claimant may not have the requisite knowledge because on the facts that could not have been obtained without the report of a relevant expert.[13]

(e) *Identity*: a claimant needs to know the identity of the defendant before knowledge is sufficient to commence the limitation period.[14] For example, this is an important consideration in asbestosis claims where ascertaining the identity of the relevant employer may be difficult over a protracted period of employment with many employers.

Case Law

There has been a plethora of cases on the issue of "knowledge of injury". The Court of Appeal has complained that limitation is overloaded with case law and so the definitive case for practitioners to consider is *Spargo* v *North Essex District Health Authority*:[15] This case set down the principles by which "knowledge of injury" is to be defined:

(a) The knowledge required to satisfy section 14(1)(b) is a broad knowledge of the essence of the causally relevant act or omission to which the injury is attributable.

---

10. Section 14(2): see *Dobbie* v *Medway Health Authority* [1994] 1 WLR 1234: *Collins* v *Tesco* [2003] LTL 24/7/03.
11. Note that a claim can be made for an exacerbation of an existing injury and the limitation period may then be different. See below.
12. *Rowbottom* v *Royal Masonic Hospital* [2003] PIQR P1.
13. *Heyes* v *Pilkington Glass Limited* [1998] PIQR 303: *Sniezek* v *Bundy (Letchworth) Limited* [2000] PIQR 213.
14. *Rush* v *JNR (SMD) Limited* [1999] LTL 8.11.99.
15. [1997] PIQR 235: See also *Forbes* v *Wandsworth HA* [1996] 7 Med LR 175. [1996] 4 All ER 881.

(b) "Attributable" in this context means "capable of being attributed to", in the sense of being a real possibility.

(c) A claimant has the requisite knowledge when he knows enough to make it reasonable for him or her to begin to investigate whether there is a case against the defendant. That is, if he so firmly believes that his condition is capable of being attributed to an act or omission which he can identify (in broad terms) that he goes to a solicitors to seek advice about compensation.

(d) There is no requisite knowledge if he thinks he knows the acts or omissions that ought to be investigated but is in fact barking up the wrong tree. Or if he is not sure that the condition is related to the defendant's conduct and he needs to check with an expert.[16]

(e) The test is subjective. What did the claimant know? Once a claimant is clear in his or her mind about the causal connection between the injury suffered and its cause (the identity of the relevant defendant having been established) then the limitation clock will probably be triggered. This principle was qualified in *O'Driscoll* v *Dudley Health Authority*,[17] where it was held that the reference to "need to check with an expert" is a need for an expert's opinion before the claimant can have any knowledge of the possibility of attributing her condition to the act or omission of the defendant.[18]

(f) If a particular injury claim is time barred but then a supervening connected injury subsequently develops then that second injury claim may not be time barred. Depending on the facts of a case, a claim may still be made for a separate exacerbation injury (e.g. pleural thickening and asbestosis).[19]

16. *Heyes* v *Pilkington Glass Ltd* [*Supra*].
17. [1998] Lloyds Rep Med 210.
18. See also *Rowbottom* v *Royal Masonic Hospital* [2003] PIQR P1.
19. *McManus* v *Mannings Marine Limited* [2001] EWCA 1668.

Constructive knowledge

Section 14(3)

Whilst a claimant may claim not to have the requisite actual knowledge within the meaning of the Act (it being a subjective test), this is balanced by section 14(3) of the Act. It provides the power for the court to impose a date of constructive knowledge on the claimant. It holds that despite a claimant's actual knowledge of the date of injury, nevertheless objectively the claimant ought to have known of the injury earlier than claimed.[20] A date of knowledge is thereby imposed by the court and it overrides actual knowledge of the claimant.[21]

Obtaining Medical Expert Evidence

First, a claimant is expected to take medical advice if in all the circumstances it is reasonable to do so. But, for example a claimant is not expected to ignore the advice of a general practitioner and seek the opinion of a consultant merely because initial reassurance is given.[22] Secondly, in many cases knowledge of the date of injury can only be imputed to a claimant on expert medical diagnosis. Until such time a claimant may not have the requisite actual or constructive knowledge, although in some circumstances the nature of the condition may make the cause of the injury obvious.[23] One issue is whether it is reasonable for a claimant to consult a medical expert. In *Rowbottom* v *Royal Masonic Hospital*,[24] it was held in the context of a clinical negligence claim that a claimant was justified in arguing he did not have the relevant knowledge for limitation purposes until he received the advice in a medical report.

20. *Hind* v *York Health Authority* [1998] PIQR 235.
21. See *Forbes* v *Wandsworth Health Authority* [1996] 7 Med LR 175. *Fenech* v *East London and City Health Authority* [2000] 1 Lloyds Rep Med 35.
22. *Penney* v *British Steel* [1997] Lawtel 1.7.97.
23. *Hind* v *York Health Authority* [1997] 8 Med Law Rep 377. *Smith* v *Leicester Health Authority* [1998] 3 All ER 324. *Jeffrey* v *CMB Specialty Packaging Limited* [*Supra*]. *Spargo* v *North Essex Health Authority* [*Supra*].
24. [2003] PIQR P1.

Practical Legal Implications

As to a claimant's "knowledge" of the date of significant injury within the Act and the practical legal implications, these are summarised below:

(a) Where a claimant claims not to have knowledge within s.14 of the Act and a defendant pleads a limitation defence, then the claimant may seek a declaration on the expiry of the limitation period.

(b) If a claimant claims that the limitation period has not expired or alternatively seeks to disapply the limitation period then the application should be for a declaration, and in the alternative an application under s.33 of the Act.

(c) Whether the application is made at an interim stage or at trial depends partly on whether the issues on limitation are bound up with those on the merits that will be heard at trial in any event.

(d) Identification of the date of knowledge of injury under the Act may require analysis of the medical notes. A claimant's history usually provides an indication of the date of knowledge of injury. If the medical notes raise a limitation point then a detailed explanation is required. Specifically, defendant practitioners ought to seek clarification under CPR Part 18 at an early stage.

(e) The relevance of an attendance on a doctor is that first it may be sufficient for a court to determine whether the injury was "significant" at that stage, depending on the facts[25] and secondly it may show that the claimant had made a connection between the injury and the relevant employment.

(f) Where a claimant denies actual knowledge of injury, there may be constructive knowledge, thereby overriding his or her actual knowledge. For example, even if a claimant is reassured there may still be grounds for making further enquiries. Full instructions are required on the point.

---

25. *Pape* v *Cumbria County Council* [1992] 1 All ER 211.

(g) The main limitation problems relating to industrial disease cases are:

- Defining the date of knowledge of the injury as, for example, respiratory diseases can lie dormant for many years.

- Assessing knowledge of injury with symptoms, some of which are explicable on constitutional bases or other non-negligent causes, (e.g. smoking).

- Fixing the date when the claimant first made the connection between the injury and the facts on which the allegations of negligence are based, particularly where there is a history of conflicting medical evidence.

- Identifying the appropriate defendant.

- Identifying the relevant statutory limitation provision, where the cause of action accrued before the commencement of the Act.

Relevant Case Law

Summarised below are cases which detail the court's approach to limitation in industrial disease cases.

Deafness

Irshad Ali v Courtaulds Textiles Limited[26]

The claimant worked in a cotton mill from 1969 to 1988. He suffered increasing deafness, particularly from 1990. But he was only told in 1991 by a community worker that it could have been caused by working in the cotton mill. He consulted a solicitor forthwith thereafter and proceedings were issued in 1995. The Court of Appeal fixed knowledge of the injury at 1991, but the knowledge of whether this was noise induced or age induced was a matter that had to be clarified by a doctor. The claim was not statute barred as the claimant's date of knowledge could only be fixed once he had expert advice.[27]

26. [1999] Lloyds Rep Med 301.
27. As to who is a relevant expert under 14(3) see *Smith* v *Central Asbestos Limited* and *Central Asbestos Co Limited* v *Dodd* [1973] AC 518. *Henderson* v *Temple Pier Co Limited* [1998] 3 All ER 324. *Hayward* v *Sharrad* [1998] 56 BMLR 155.

Jeffrey v CMB Speciality Packaging (UK) Limited[28]

The claimant was exposed to high levels of noise in his work place and suffered industrial deafness in 1969. No proceedings were issued until 1993. His reason for the delay was that he did not make a connection between his deafness and his working conditions until 1992. On appeal, the issue was the terms of investigations that a claimant should undertake. It was held:

(a) The test applied was unrealistic in expecting the claimant to make such enquiries.

(b) The claimant had been given a natural cause by his consultant in 1970.

(c) It went beyond the requirements of s.14(3) to make such enquiries of nurses at his place of work about the cause of his tinnitus, particularly as his employers were not taking any steps to provide him with protection.[29]

Milner v Hepworth Heating Limited[30]

The claimant suffered industrial deafness as a result of equipment used at the defendant's premises. A writ was issued in 1995 and the point was whether he had knowledge of his injury prior to 1992. He had started work in 1962 with the defendant and hearing protection was made available in the 1970s. The claimant only began to wear hearing protection in 1982 and he noted his hearing deteriorated in 1990, but he had wax problems with his ears. He was seen by a consultant in 1993. The defendant's appeal was dismissed as the Court of Appeal held that the section 11/14 issues were a matter for the judge on the facts. Further the perception of deafness was a subjective matter and it was common for people with this difficulty not to appreciate it. On constructive knowledge the claimant could not have obtained relevant knowledge at an earlier date.

28. [2000] C.P. Rep 1: The issue of not having knowledge without first having a medical report in the appropriate case was upheld in *Heyes* v *Pilkington Glass* [1998] PIQR 303: a case concerning vibration white finger.

29. Note that refusing medical tests can result in a claimant being fixed with knowledge at that date: *Denford* v *Redbridge and Waltham Forest Health Authority* [1996] 7 Med Law Rep 376.

30. [1998] Lawtel 28.7.98.

Vibration White Finger

Heyes v Pilkington Glass Ltd[31]

The claimant was employed as a crane operator and he suffered vibration induced white finger/hand-arm vibration syndrome ("VWF"), in the course of his employment. Liability was established at first instance. The claimant had suffered from VWF from 1980 until his employment was terminated in 1987. He consulted solicitors in 1990 and proceedings were commenced in 1993. On appeal it was held that the claimant did not have the requisite knowledge required by s.11 or s.14(1) of the Act to issue proceedings until he had obtained a medical report. Only then did he know his condition was occupationally induced. Further, there was no evidence that the claimant knew or ought to have known of the issue of VWF.

McManus v Mannings Marine Limited[32]

On a trial of the preliminary issue of limitation the claimant's claim was dismissed at first instance. He had suffered VWF having worked in the ship building industry since 1968 and for the defendant since 1989. He claimed from the DSS in 1992 relating to his VWF, which was handled through alternative dispute resolution. He was seen by a consultant in 1993 and employed again by the defendant from 1993 to 1999. Proceedings were issued in 1999. At first instance knowledge under section 14 was stated to be in 1992, when the application to the DSS was made. The claimant contended the defendant was liable for the exacerbation, which the claimant did not have knowledge of until he suffered a deterioration. The Court of Appeal upheld the claim as "the injury in question" under section 14 meant the injury for which the claim was brought. The claimant did not have knowledge of the extent of the exacerbation.

Asbestosis

Shuttleton v Duncan Stewart & Co Ltd[33]

The claimant made five claims against his former employers relating to his employment between 1930 and 1989. He suffered from pleural plaques, pleural thickening and asbestosis,

31. [1998] PIQR P303.
32. [2001] EWCA Civ 1668.
33. [1996] SLT 517.

allegedly due to asbestos exposure. He underwent tests in relation to breathlessness in 1977 but these were normal. Medical advice was that he suffered from pleural plaques, but not asbestosis. In 1983 he was diagnosed with obstructive airways disease, and he was not diagnosed with asbestosis until 1987. The defendant alleged knowledge of the injury by 1983, when he had pleural thickening. It was held that there was no knowledge prior to 1987 such that proceedings should have been issued. Moreover, pleural plaques, pleural thickening and asbestosis were distinct conditions and so each injury would have a separate limitation period. Accordingly the asbestosis claim was not time barred.

Summary and Practice Points

The schedule below summarises the main points on this chapter, and a precis of the practical points arising on limitation disputes.

Issue	Comment
Limitation period for cases involving death or personal injury is 3 years from accrual of cause of action or knowledge of significant injury. If the claim is time barred then an application is made under s.33.	1. Ss.11 and 12 set out the criteria. 2. It is 3 years from death of the deceased, as long as the limitation period had not expired at death. 3. If the claim is statute barred then apply under s.33 to disapply it. 4. Consider whether to have an application heard at trial or as an interim issue.
• *Section 14*: defines "knowledge of a significant injury". The defendant has to be identified. • Knowledge depends on the connection between the symptoms and the system of work (attributability). Knowledge of the alleged negligence is not relevant. • *Spargo* v *NE Essex Health Authority*: sets out the four principles of knowledge of injury.	• Once the limitation period commences it cannot be stopped. • The knowledge required is of a "significant" injury. The essential issue is the first date of knowledge of the connection between the system of work and the symptoms. • Discrete limitation periods can relate to individual injuries and this includes exacerbation. The *Spargo* case is a good starting point.
Constructive Knowledge (s.14(3)) 1. Actual knowledge can be overridden by constructive knowledge (i.e. when the claimant ought to have known of a significant injury). 2. Limitation will run from the date of constructive knowledge. 3. S.14(3) expressly refers to obtaining "expert advice". 4. A claimant may have constructive knowledge long before actual knowledge (e.g. on the GP records). 5. A claimant should not ignore the opportunity to take medical advice or ignore it when given, as this may be considered as constructive knowledge.	1. A claimant's actual knowledge can be undermined by s.14(3). Practitioners must consider if claimant has taken medical advice. 2. Consulting a GP may be sufficient to obtain knowledge of injury but the explanation of the claimant is critical. Reassurance by doctors can negate knowledge, if the limitation clock has not yet started. 3. Full statements and chronologies are needed. Consideration of Benefits' Agency or other records is essential.

Issue	Comment
Practice points 1. The date of knowledge fixes the commencement of the limitation period and so indicates the period of delay. 2. A detailed statement has to be taken from the claimant to assess when the connection was made between the system of work and the injury. 3. If the claim is statute barred consider a s.33 application, heard at trial or at an interim stage. This depends on the issues and the merits of the application. Probably sooner rather than later. 4. Medical evidence may be an important factor in determining knowledge of injury. E.g. could the claimant have identified the nature of the condition at all or at a certain date in the absence of medical expert advice?	

CHAPTER 5
LIMITATION IN INDUSTRIAL DISEASE LITIGATION: SECTION 33, THE LIMITATION ACT 1980

An Overview

Limitation is a defence to an action and so needs to be pleaded. In a significant proportion of industrial disease claims a claimant will argue that he did not have the requisite knowledge of injury under section 14 to start the limitation period. This issue is dealt with in chapter 4. In contrast, if it is accepted that the claim is statute barred then the court may disapply the limitation period under section 33 of the Limitation Act 1980 ("the Act").

The section states as follows:

> 33 "(1) If it appears to the court that it would be equitable to allow an action to proceed having regard to the degree to which–
>
> (a) the provisions of section 11 or 11A or 12 of this Act prejudice the plaintiff or any person whom he represents and:
>
> (b) any decision of the court under this subsection would prejudice the defendant or any person whom he represents;

the court may direct that those provisions shall not apply to the action, or shall not apply to any specified cause of action to which the action relates."

Procedure under Section 33

The following points are to be noted in relation to section 33 of the Act:

(a) The court has a wide discretion and account is taken of all the circumstances of the case in deciding whether it would be equitable to disapply the limitation period.

(b) A balance is to be struck between the respective prejudice to the parties, namely the loss of the limitation defence by the defendant and the loss of the right to bring a claim by the claimant if the limitation period is not disallowed.[1] It is not a balance of hardship but one assessed according to the application of the factors in the section.[2]

(c) The Court of Appeal rarely interferes on findings of fact given the wide discretion of the judge. Accordingly on appeal it has to be shown that the trial judge took into account factors that should have been ignored (or *vice versa*) or was palpably wrong.

(d) The discretion under section 33 is exercised according to the six factors set out in section 33(3).[3]

(e) The burden of proof is on the claimant.

1. *Long* v *Tolchard and Sons Ltd* [2001] PIQR P.18.
2. *Thompson* v *Brown Construction (Ebbw Vale) Limited* [1981] 2 All E.R. 296 HL.
3. As set out in *Farthing* v *NE Essex Health Authority* [1998] 2 Lloyds Rep. Med 37. *Burke* v *Ashe Construction Ltd* [2003] EWCA Civ. 717.

The factors under S.33(3)

This sub-section provides the rubric for the practitioner to assess the merits of the application. They are exemplary but not definitive.[4] The six factors are as follows:

1. Length and reasons

Subsection 33(3)(a): *the length and reasons of the delay by the claimant*

- The test is subjective. The claimant has to put forward the reasons for delay and the court assesses those reasons. The relevant delay under this section is the delay from the expiry of the limitation period to date of issue, but the total period of delay can also be taken into account.[5]

- The claimant has to explain and not seek to "excuse" the period of delay. Claimants have to explain the delay in each relevant period and a detailed statement is essential.

- The period of delay is a matter of fact assessed by the court. The longer the delay the more detailed the explanation to be given by the claimant.[6]

- The culpability of the claimant in the delay is an important factor and ignorance of the law is a material factor under this section.[7] Moreover the fact that the claimant was a non-litigious person and fearful of losing his or her job can also be relevant factors.[8]

- The receipt of negative medical advice is also an important factor.[9]

4. *Barrand* v *British Cellophane Plc: The Times* 16.2.95.
5. It is very important to issue proceedings to stop the "clock ticking" as failure to do so only adds to the delay.
6. As to a short delay: *Steeds* v *Peverel Management Services Limited* [2001] EWCA Civ 419 where the delay was 49 days. *Skerratt* v *Linfax Ltd* [2003] LTL 6/5/2003.
7. *Berry* v *Calderdale MBC* [1998] Lloyds Rep Med 179. On ignorance of the law see *Halford* v *Brookes* [1991] 3 All ER 559.
8. *McCafferty* v *Metropolitan Police District Receiver* [1977] 2 All ER 756.
9. *Smith* v *National Health Service Litigation Authority* [2001] Lloyds Rep Med 90.

2. Cogency

Subsection 33(3)(b). *The effect on the cogency of the evidence by reason of the delay*

- "Cogency" is defined in terms of the ability of the defendant to defend the claim when there has been a delay. This may have affected the quality of the evidence on which it would choose to rely if the claim had been brought within the limitation period. The relevant period is from the end of the limitation period to the date of issue of proceedings.[10]

- It is important first, to assess what the relevant evidence is. This is usually witness and documentary evidence, but the categories of evidence are not limited. Secondly, assessment needs to be made of what witnesses and documentary evidence would have been available if the claim was brought within the limitation period, as on the facts of a case it might not make any difference (e.g. if a witness died within the original limitation period).

- This section is particularly relevant to industrial disease claims where there can be a delay from the relevant period of employment to the date of onset of the symptoms. Factories may have closed and former employees may be untraceable. It is very important to assess exactly whether and to what extent the evidence has been adversely affected by the passing of time.

Cogency: Points from a Claimant's Perspective

From a claimant's perspective the following points should be addressed:

(a) Identify the system of document retention and/or contact with former employees that the defendant administered. If it had no system at all then the loss/destruction of documentation may have taken place during the limitation period in any event.

10. See *Price* v *United Engineering Steels Limited* [1998] PIQR 407 as to whether loss of cogency can be inferred.

(b) Identify the documents and witnesses potentially relied on by both parties. Are there still sufficient witnesses who can give evidence on the system of work? If not, why not?

(c) Identify the core documents required in this type of case and make enquiries as to what has happened to them in the intervening period. Arguably the loss of documentation may not affect the cogency of the evidence in any event.

(d) If witnesses and/or documents have been lost, then obtain a clear explanation of the attempts to find/trace them. Then assess how in practice this affects the ability of the defendant to present its defence.

(e) Consider whether medical or other expert evidence can assist on the issue of cogency of the evidence without the need for detailed oral evidence.

Cogency: Points from a defendant's perspective

From a defendant practitioner's perspective:

(a) Assess and identify the system for the retention of documentation. For example was it company policy that insurers would destroy the relevant file after a certain period? If so, then obtain evidence from the relevant witnesses.

(b) Having shown there is a system in place then analyse what effect the passing of time has had on (1) witnesses and (2) documents (including whether the defendant is still in business).

(c) Provide evidence from witnesses who can speak with knowledge on the point (as opposed to an inexperienced witness with no knowledge of the case, the issues or the evidence). Where necessary, show what attempts have been made to trace witnesses and/or documents.

(d) Decide whether the case can be argued easily on extant documentation and so the fact that there are no live witnesses is irrelevant (or *vice versa*).

(e) Consider obtaining medical or other expert evidence to prove that the passing of time has affected the ability of the defendant to defend the case on liability.

(f) Analyse all the medical records especially those relating to any application for benefits (e.g. industrial disablement benefit and the related medical examinations and application forms), which could show there has been an unreasonable delay since the date of knowledge and/or other inconsistencies with the claimant's version of events.

3. Conduct of the defendant

Subsection 33(3)(c) *the conduct of the defendant in the relevant period*

This subsection relates to the conduct of the defendant in complying with its duties in the litigation process. It does not cover the acts of insurers. The conduct has to be intentionally reprehensible as opposed to incompetent. If so then the court can take this into account in deciding whether to disapply the limitation period.[11] The fact the defendant has made interim payments to this or other claimants does not preclude it from resiling from its admission on liability.

4. Period of disability

Subsection 33(3)(d): *the period of disability suffered by the claimant*

A claimant may be under a disability, as defined by section 28 of the Act.[12] In some cases a claimant could be so badly injured that he may not have the capacity to understand the proceedings and so had no knowledge of the accrual of the cause of action or insufficient knowledge of the same. This may be relevant to the "knowledge of injury" under s. 14, but also to s. 33. Further, if there is a period of disability of the claimant which ends then the court can take this into account in deciding on the period and the reasons for the delay in issuing proceedings.

---

11. *Drury* v *Grimsby Health Authority* [1997] 8 Med LR 38: As to ADR schemes which are on-going and limitation see: *Bytheway* v *British Steel Plc* (1997) LTL 26/6/97.
12. *Thomas* v *Plaistow* [1997] PIQR 540.

5. Prompt and reasonable action?

Subsection 33(3)(e) *The extent to which the claimant acted promptly and reasonably once he knew whether or not the act or omission of the defendant to which the injury was attributable might be capable at that time of giving rise to an action for damages*

Once the claimant knows of the cause of action in negligence then he has to progress the claim without delay. To the extent the claimant fails to do so then this can be fatal to the exercise of the court's discretion in his or her favour. "Prompt and reasonable" is not precisely defined but will be construed in a wide context and starts from the date the claimant knows of the cause of action.[13] If the claimant has not so acted then a complete explanation is required.[14]

6. Obtaining medical or legal advice

Subsection (33(3)(f): *the steps, if any taken by the claimant to obtain medical or legal advice and the nature of such advice received*

A claimant does have to make reasonable efforts to make relevant enquiries as to the cause of his or her condition and who is responsible for it. Negative medical or legal advice can be a legitimate reason to explain the delay,[15] but it is important that a claimant does not delay in obtaining such advice, whatever the quality.[16] This section may require the waiver of privilege by a claimant on reports and legal advices and a careful assessment of the nature and extent of that proposed course of action is required.

Additional points

(a) Ignorance of one's legal rights can be a reason which the court takes into account in exercising its discretion in this section.[17]

13. *Skitt* v *Khan* [1997] 8 Med LR 105.
14. *Berry* v *Calderdale* [1998] 5 Lloyds Rep Med 179.
15. *Das* v *Ganju* [1999] Lloyd's Red Med 198.
16. *Skitt* v *Khan* [*supra*].
17. *Halford* v *Brookes* [1991] 1 WLR 428: *Johnson* v *Busfield* [1998] 38 BMLR 29. As to where proceedings have been started and then discontinued within the limitation period with a later application under s.33: see: *Young* v *Western Power Distribution Plc* [2003] EWCA 1034.

(b) If the fault for the delay has been caused by the claimant's solicitors then this should not be held against the claimant.[18] The point is illustrated by *Corbin* v *Penfold Metalising Co Limited*.[19] The claimant was employed as a metal sprayer and he suffered a chest condition and so saw a specialist in March 1993. Having been told he had iron deposits in his lungs probably caused by his employment he instructed solicitors. No proceedings were issued until August 1996. On appeal against the strike out of his claim it was held that there was no rule of law which attributed the negligence of the solicitors to the claimant, in the absence of a rule of law expressly making a client responsible for the actions of the solicitor. The claimant was not culpable and so the faults of the solicitor should not be attributed to him.

(c) The merits of the claim are an important aspect in the exercise of the discretion: In *Bates* v *Leicester Health Authority*[20] it was stated

> "..it is always prejudicial to a defendant to be deprived of a limitation defence, but it is less inequitable or unfair where, as here, the plaintiff has a strong case than where he has a weak one."[21]

Practice points

In practice practitioners need to consider these points:

(a) Whether there was knowledge of injury within section 14 so that the limitation period has not in fact expired, or whether the point is conceded and the claimant relies solely on section 33 of the Act.[22]

(b) The reasons for delay are many and varied, but it is vital to have all the explanations contained within a claimant's statement. Particulars are required relating to the delay post expiry of the limitation period, but also any further delay from the accrual of the cause of action. Delay in the

18. *Das* v *Ganju* [*supra*].
19. [2000] Lloyd's Rep Med 247.
20. [1998] Lloyds Rep Med 93.
21. [1998] 3 Lloyds Rep Med 93, relying on *Forbes* v *Wandsworth Health Authority* [1996] 3 WLR 1108 at 1120G.
22. The applications can be made in the alternative.

limitation period itself can affect the exercise of the court's discretion as part of the circumstances of the case.

(c) Chronologies are useful tools to highlight the periods of delay in an individual case and so show up the strengths and weaknesses on this point.

Procedure for a Claimant's Application

Under the Civil Procedure Rules an application by a claimant is made by a Part 23 application notice, setting out in detail the principal relief sought. A defendant may apply under CPR Part 3 for an order under the Court's own initiative, under rule 3.3 or, alternatively, under rule 3.4. Alternatively it can apply under Part 24 for summary judgment on the limitation issue under Part 24.2(a)(i), namely that, the claimant has no real prospect of succeeding on the issue.

An application can made as a preliminary point at an interim stage, or at trial. Careful consideration has to be given whether to take the claim to trial with a live limitation point, particularly if the claimant's case is weak on this point. Conversely, where the claimant's case is strong on the limitation point then the claimant's solicitors can leave the decision on an application up to the defendant's solicitors.[23] An important factor is whether the facts and issues on the merits are bound up with the limitation issue, so that the point is better taken at trial.[24]

The Defendant Company is Dissolved

In principle, a company which has been dissolved can be restored to the companies register under s.651 of the Companies Act 1985, for the purpose of a personal injury action. This is necessary to obtain a judgment that can be enforced against the relevant insurers. But an application under s.33 may be required if the claim is statute-barred. The court will consider its discretion under s.33 when exercising its power under s.651 of the 1985 Act. In *Harrison* v *Sun Alliance and London Assurance Plc*,[25] the claimant was injured at work and claimed against the employer company. Before judgment the company went into liquidation and was dissolved, but the claimant did not know of this development. He applied to

23. As to the substitution of parties and claims after proceedings are issued and once the limitation period has expired see CPR Part 19.
24. *Roberts* v *Winbow* [1999] Lloyd's Rep Med 31.
25. [1998] CLY 517.

reinstate the company in order to commence new proceedings. It was held the fact that the claimant did not know the company had been dissolved was a ground under s.33 in the claimant's favour.[26] Fresh future proceedings would have reasonable prospects of success, particularly as the company had been dissolved without making proper provision in relation to his claim. Furthermore, an application under s.33 is strengthened if the limitation period had not expired before the dissolution of the company.[27]

It is therefore important to issue proceedings as soon as possible, preferably having previously notified the defendant's solicitors or insurers of the potential claim.[28] In *Long* v *Tolchard and Sons Limited*[29] it was held that a claimant must disclose all relevant facts and circumstances at the hearing of the application. If subsequently it became apparent that a claimant had failed to give a truthful and accurate history of the injury then an order under s.33 in his or her favour could be set aside.

A Practical Approach from a Defendant's Perspective

(a) If a limitation point is raised it must be decided whether to take it at an interim stage or at trial. If the case is vulnerable on the merits then there could be proportionate costs savings in taking the point at an early stage.

(b) Statements have to be carefully drafted and the rubric provided by section 33(3) should be adopted. Note that in many cases a series of witness statements may be required.

(c) Set out a chronology that highlights the delay and seek further information under CPR Part 18 on the claimant's perceived weak points on the claim.[30] This precludes a later and exculpatory explanation of the delay at the hearing if the Claimant has provided a previous inconsistent statement.

(d) As to cogency, the case has to be clearly set out on the documents and the witness evidence. Detailed evidence from a witness with knowledge is required. It also has to be shown there was some system for the retention of

26. *Re: Philip Powis Limited* [1998] 1 B.C.L.C. 440.
27. *Re Workvale Limited (In Dissolution)* [1992] 1 WLR 416.
28. Part 23.2(4) CPR expressly provides for applications prior to the issue of proceedings.
29. [2001] PIQR P.2.
30. Obtain all the relevant medical records (including DSS and Benefits Agency records).

documents, or that the business in question has long since been dissolved and all documents destroyed. Would expert evidence help make the cogency point further? Can witnesses be traced or contacted through a video link or a statement be obtained on a hearsay basis?

(e) If the claimant is relying on incorrect medical or legal advice then seek a copy of the file and/or report as it may be material in this or other respects to the application.

(f) If the application is made at an interim stage then ensure the claimant attends for cross examination.

Case Law on Limitation S.33

The cases in this section provide an indication of the court's approach to the exercise of its discretion under section 33 in the context of industrial disease and injury litigation.

Mesothelioma

Smith v White Knight Laundry Ltd [2002] 1 W.L.R. 616

The claimant was the widow and personal representative of a man who had died from mesothelioma arising, allegedly, from his exposure to asbestos in the course of his work. The defendant appealed against a decision setting aside a direction under the Companies Act 1985 s.651 relating to its dissolution and restoration to the register for the purpose of the limitation period. The cause of action accrued, by virtue of the restoration order, on the date on which it would have accrued but for the dissolution. The period between dissolution and restoration was irrelevant for limitation purposes.

Paterson v George Wimpey & Co Ltd [1999] S.L.T. 577

A claim was brought by the widow, executrix and children of the claimant/deceased who died from mesothelioma caused by exposure to asbestos dust while employed by the defendant between 1948 and 1951. The defendant asserted the action was time barred. It was held that the claimant succeeded and it was for the defendant to prove when the injury occurred and when the limitation period was triggered. No limitation period could run before injury was caused and the fact that the claimant was not aware of his injury was irrelevant.

Arnold v Central Electricity Generating Board [1987] 3 W.L.R. 1009

In a mesothelioma case the deceased had been employed from 1938 to 1943 by the defendant as a boiler cleaner. In 1981 he was diagnosed as suffering from mesothelioma; he died in 1982. The defendant claimed the action was time barred by the Limitation Act 1939. The Court of Appeal held that where a statute was clearly intended to have some retrospective application, it had to be strictly confined in scope to give effect only to what was expressed in clear language. The defendants were therefore entitled to rely on their rights and accrued defence.

Asbestosis

Cowan v Toffolo Jackson & Co Ltd [1998] S.L.T. 1000

The claimant sued his former employers when he was diagnosed as suffering from asbestosis. Between 1943 and 1956 the claimant was employed applying ceramic tiles on ships' hulls in various shipyards and he was working with asbestos. From 1956 until 1986 he worked for the defendant carrying out the same duties in the same circumstances. He was advised on retirement in 1986 by his medical advisers that he did not have an asbestos related disease. A further x-ray in 1991 showed pleural plaques on both sides consistent with previous exposure to asbestos. An expert in 1992 confirmed the asbestosis and proceedings were issued in May 1992. The limitation point was raised and reliance was placed on the claimant's known breathlessness in 1986. The claimant contended that he had been medically advised that he was not suffering from asbestosis at the time and so did not follow it up. It was held that the claimant's knowledge of the asbestos exposure and the conditions at work meant he did have knowledge of his condition and injury such that the claim was time barred. As he had not addressed the issue of his medical history then the court would not exercise its discretion.

Shuttleton v Duncan Stewart & Co Ltd 1996 S.L.T. 517

Five claimants brought actions against their employers relating to pleural plaques and asbestosis caused by exposure to asbestos in their employment in the period 1939 to 1980. Further relevant facts were:

(1) One claimant had smoked about 30 cigarettes per day, and developed breathlessness. A lung biopsy in December 1977 had shown no definite abnormality and he was re-assured that his condition related to smoking.

(2) On that basis an appeal to a medical tribunal was withdrawn. Further tests in 1983 showed obstructive airways disease, but he was advised this was unrelated.

(3) In 1986 the claimant applied for disablement benefit on the basis of pleural thickening, but the application had been refused. He claimed he did not know of the asbestosis until 1987.

It was held that "sufficiently serious" meant an injury that was more than just fractionally above the cut off for damages. By 1983, pleural thickening was diagnosed but the claimant was reassured about his condition and there was no basis for him to claim at that time. The claimant had been aware prior to 1987 that his condition was sufficiently serious to justify bringing an action of damages. It was also noted that as pleural plaques, pleural thickening and asbestosis were separate diseases the limitation period in relation to each could be different. Even if a claim based on pleural plaques was time barred the claim in relation to asbestosis (which caused all the damage) could proceed under the equivalent of s.33.[31]

McLaren v Harland and Wolff [1991] S.L.T. 85

A claimant was employed with a firm of shipbuilders between 1953 and 1960. In 1983 he was diagnosed as suffering from asbestosis and claimed benefit from the DHSS. He died in 1988 and thereafter an action was brought against the employer claiming his death was caused by asbestosis. It was held the deceased knew he had the disease in or about 1983 and further that it was connected to his working conditions. On that basis

31. See also: *Carnegie* v *Lord Advocate [No.3]* [2001] SC 802.

the action was statute barred but it was equitable to permit the claim to proceed as the deceased did not know of his right of action in his lifetime.

Respiratory injury

Corbin v Penfold Metallising Co Ltd [2000] Lloyd's Rep. Med. 247

The claimant was formerly employed as a metal sprayer and was advised in 1993 by a chest specialist that he was suffering from a condition caused by iron deposits in his lungs, which was caused by his working conditions. The claimant instructed solicitors but a writ was not issued until August 1996. The claim was struck out as the limitation period had expired. On appeal the claimant asserted the delay was caused by his solicitors and the appeal was allowed. It was held there was no rule of law that the fault of solicitors should be attributed to the claimant, unless the claimant was so bound as a matter of law. This is particularly the case where the claimant was not at fault in pursuing the claim.

Back Injury

Long v Tolchard & Sons Ltd [2001] P.I.Q.R. P2

The claimant brought an action for damages for personal injury in 1990 relating to a back injury at work that was caused in 1983. He claimed he had no back problems prior to 1983. The limitation issue was found in the claimant's favour and liability was established. However at a further trial evidence showed that the claimant had not made full and frank disclosure of his previous back problems before 1983. On the defendant's appeal it was held that the discretion to disapply meant the court considered all the circumstances of the case. As there was now an additional four year period of delay that was unexplained and which had not been disclosed then the claim was statute barred.

Hearing loss

McAloon v Lord Advocate [1999] G.W.D. 1393

The claimant brought a case against the defendant for damages for significant hearing loss sustained during the course of army service, in particular through mortar fire, between 1986 and 1989. He was discharged in 1990; the action was brought in 1997. The claimant first complained of hearing loss in 1989, and was referred for an audiogram which was suspicious of deafness. By 1990 he thought his deafness had been caused by his service. In 1995 the claimant consulted solicitors. He said he had not acted before on account of ignorance of what to do. The action was dismissed, as it was reasonable for the claimant to have become aware of all the facts within a reasonable time. He was not misled into letting time go by and the defendant was prejudiced in defending the claim.

Price v United Engineering Steels Ltd (1998) PIQR 407

P, a former steelworker, worked with D1 from 1961 to May 1964, and with D2 from 1969 to 1980. He was aware of the onset of deafness in the late 1970s and was examined by a works' doctor in 1978. No ear protection was ever provided. In 1992 he issued proceedings against D1 and D2 claiming damages for his hearing loss. At first instance it was held that the primary limitation period expired in 1989. On appeal it was held that:

(a) in a noise induced hearing loss case, the fact that other claims had been settled by insurers did not mean there was no prejudice to the insurers in this case.

(b) There was a wide discretion under s.33 and the fact the works had now been closed for a long period and that records were destroyed and witnesses untraceable entailed prejudice.

(c) The fact that a defendant failed to call sufficient evidence on the cogency of the evidence did not mean that a judge could not infer the same.

Nimmo v British Railways Board [1999] S.L.T. 778

The claimant sought damages from his former employers for the loss of his hearing, from exposure to excessive levels of noise at his work prior to 1985. The claimant knew of the hearing problem in 1977 and medical advice was given in 1992. No proceedings were issued until 1996, at which time the claimant saw another doctor. Proceedings were started in 1997. It was held that the claim was dismissed, as the claimant could have commenced the claim by 1995. Further the fact that the defendant faced a number of similar claims was not a reason to permit the claim to proceed.

Hand-Arm Vibration Syndrome

Allen v British Rail Engineering Limited [2001] PIQR Q.101

The claimant sustained a HAVS injury in the course of his employment with the defendant. His symptoms began in the 1960s and worsened after 1973. The claim was time barred by 1992. It was held that in the absence of culpability on the part of the claimant there was no reason why the court should not exercise its discretion in his favour. He was not aware of his claim and had limited understanding of the legal complexities. He was also a stoical man who did not want to trouble his doctors.

Industrial Dermatitis

Crawford v Deaduck Limited (In Liquidation) [1997] QBD Nelson J: Unreported

The claimant was a floor layer who contracted dermatitis in the course of his employment from 1989 to 1992. The cause of action accrued in 1989 and the claimant first consulted solicitors in 1992. A writ was issued in 1995. The claimant claimed he did not have the relevant knowledge of injury under s.14 until 1992 when a consultant dermatologist confirmed his allergy to epoxy resins. It was held that consulting solicitors did not mean that a claimant had knowledge of the injury and each case is considered on its facts. But the claimant suspected a connection between the glue used at work and his symptoms shortly after they presented. He actually purchased cotton gloves for this reason and he knew when he consulted his solicitors that the medical evidence confirmed his suspicions as regards the dermatitis. Limitation therefore expired in 1995, but as the case was strong and the delay short then the discretion was exercised in his favour under s.33.

Summary of the Chapter

Point	Comment
1. Section 33 applies where the claim is statute barred. It provides a mechanism to disallow the limitation period. 2. The power is equitable and discretionary and based on the balance of prejudice between the parties.	Practitioners ought first to ensure whether in fact the claim is statute barred, particularly in the context of s.14 knowledge of injury.
The main provisions are contained in s.33(3) and these provide a guide to assess whether the limitation period should be disapplied.	Subsection 33(3) should be used as the basis of the claimant's or defendant's statements.
The main factors under s.33(3) are • An explanation of the delay. • The effect on the cogency of the evidence of delay. • What actions the claimant took once he knew there was a cause of action.	• Explanations of delay must be given and not excuses. The standard is subjective and so any explanation can be sufficient. • Cogency of evidence should be considered in terms of documents and witness evidence. • A claimant must not delay once the cause of action is known as this will invariably be fatal to the application.
Additional points: • Ignorance of the law can be an explanation. • The fault of the solicitors should not be imposed on the claimant. • The merits of the claim are important as the stronger the merits the more reason to disapply the limitation period. • The approach overall is to consider the balance of prejudice.	• Consideration of when to make the application: at trial or at an interim stage. The weaker the case the stronger the reason to take it at an interim stage. • There is nothing preventing a defendant applying to strike out the claim. • Consideration has to be given whether to call the claimant or other witnesses. • Defendants may want the claimant called in any event. • Medical evidence may be required to explain certain aspects of the application.

CHAPTER 6
ASBESTOS RELATED DISEASE CLAIMS

Section 1: The Legal Perspective

Overview

Many claims for damages for industrial respiratory diseases relate to the exposure of employees to harmful substances in the workplace. This will include respiratory diseases arising in miners, steel workers and others working in heavy industry. Whilst reference is made to claimants who have suffered respiratory diseases through dust inhalation (e.g. miners) the main emphasis of this chapter is on asbestos-related disease litigation. Asbestosis is a slowly progressive condition with an insidious onset, which may cause no symptoms in the early stages. Mesothelioma may arise decades after exposure to asbestos, due to its long latent period. The various medical conditions caused by asbestos exposure have been known about for many years.

Claims are complex as many of the employers will have ceased to exist at the time of the onset of the claimant's symptoms. Further, asbestos related claims combine problems over limitation, breach of duty and causation. The assessment of damages is also invariably contentious, not least as to whether provisional damages are appropriate. In recognition of the scale and complexity of the problem a statutory compensation scheme was initiated under the Pneumoconiosis (Worker's Compensation) Act 1979.[1] This compensated victims of certain respiratory diseases at work, where the defendant company had been dissolved. The compensation was modest and had to be set off against any award made in subsequent litigation.[2]

1. Based on *Griffiths* v *British Coal* [1998] Lawtel 23.1.98: the government inherited liability from British Coal.
2. *Ballantine* v *Newhalls Insulation Co Limited* [2001] ICR 25: applying the deduction of "collateral benefits" set out in *Hodgson* v *Trapp* [1988] 3 WLR 1281 AC 807.

Additional schemes include a government scheme, which has compensated miners who have suffered respiratory diseases caused whilst in the employment of British Coal. It was commenced on 26th of March 1999 and awards were up to £50,000, with a reduction to take account of smoking histories and other causal factors. The scheme was intended to cover about 65,000 miners who were claiming compensation for injuries sustained in the course of their employment with British Coal.

Practitioners therefore need a thorough understanding of the issues in this area of personal injury litigation in the United Kingdom.[3]

The Duty of Care

An employer owes a duty of care at common law and pursuant to statute to protect employees from the foreseeable risk of harm in the course of their employment.[4] The standard of care is that of the reasonable employer, but in the context of asbestos related claims a series of statutory duties modifies the duty at common law.[5] Assuming that a duty of care is imposed by reason of the employer/employee relationship, a claimant has to prove that the system of work was negligent in that it created a reasonably foreseeable risk of injury. First, the duty of care is subject to proof that the employer had, or ought to have had, knowledge of the relevant risk of danger to health at the material time. Secondly, if the knowledge of the employer is satisfied then a claimant must still prove that it failed to take the necessary and reasonable steps to prevent injury and that it was this breach of duty that caused the injury.

Breach of Duty

The date of knowledge of the employer

In asbestos related litigation the date of knowledge of the employer varies with the increase in knowledge of the dangers

3. This chapter is restricted to claims in the UK, but as to foreign claims, *forum non conveniens* and related matters see: *Lubbe et al* v *Cape Plc* [2000] 1 WLR 1545 and the European Court judgment in *Group Josi Reinsurance Company SA* v *General Insurance Company* [2000] ILPr. 549.
4. *Page* v *Smith* [1996] 1 AC at 190. [1995] 2 All ER 736.
5. Note that statutory regulation is not subject to foreseeability, unless specifically provided for in the statute. Reference should be made to the chapter 1 for the general principles on duty of care of the employer.

of asbestos over time.[6] But a reasonable point at which to attribute knowledge of the risk of injury is 1932 onwards, based on the Asbestos Industry Regulations 1931.[7] This date should suffice for most current cases.[8] By that date, an employer's knowledge of the risk of injury to the employee imposed on it a duty to warn or otherwise protect its employees. The regulations applied to industrial processes relating to asbestos and were of industry-wide application (i.e. factories and workshops). For example, in *Bryce* v *Swan Hunter Group Limited,*[9] a ship painter died from malignant mesothelioma of the pleura. An action was brought against his previous employers, alleging his death was caused by the inhalation of asbestos dust. The defendant denied any breach of duty but liability was established, based on the finding that the risks of asbestosis were known from 1947.

The link between asbestos and mesothelioma (as opposed to asbestosis) was not known until the 1960s, but the risks of lung disease were known in the 1950s.[10] In more recent cases the employer's date of knowledge will probably not be in issue,[11] but insofar as a defendant's knowledge is an issue the following documentation is relevant in proving foreseeability of injury:

(1) The "Merewether report": E.R.A Merewether and C.W. Price *'Report on the effects of asbestos dust on the lungs and dust suppression in the asbestos industry: 1930'* [HMSO 1930].

(2) Wood and Gloyne *'Pulmonary asbestosis: a review of one hundred cases'*: [1934]

6. Note that in *Margereson* v *JW Roberts Limited* [*supra*] a factory owner was liable for asbestos-related disease caused to persons living near to it although not employed by it. The duty therefore extended "beyond the factory wall".
7. *Barclays Bank plc* v *Fairclough Building (No.2)* [1995] PIQR 152: *Margereson* v *JW Roberts Limited* [1996] PIQR P.358 where it was held the date of knowledge of the dangers presented by asbestos pre-dated 1933 when the 1931 regs came into force.
8. But see *Smith* v *P and O Bulk Shipping Limited* [1998] 2 Lloyds Rep 81, where it was held that a reasonable shipowner would only have known by 1971 of the dangers to dock workers who did not come directly into contact with asbestos but were at risk.
9. [1987] 2 Lloyds Rep 426.
10. *Shell Tankers UK Limited* v *Jeromson and Cherry Tree Co Limited* [2001] EWCA Civ 101. [2001] PIQR P.19.
11. See also *Hancock* v *JW Roberts Limited* (1996) PIQR 154 which related to claimants who suffered mesothelioma, and who were resident and had gone to school near the defendant's factory which processed asbestos. Liability was established on the basis that there was actual or constructive knowledge of the risk of emissions from the factory beginning from 1933.

(3) The annual reports of the Chief Inspector of Factories 1931–1956 and thereafter to 1982.

(4) "Mortality from lung cancer in asbestos workers" [1954] *British Journal of Industrial Medicine.*

(5) A series of widely published warnings contained in published industrial health journals in the period from 1954 to 1982.[12]

Breach of Duty

Statutory Duty

A series of statutory regulations control the use of asbestos in the workplace. A summary of the most important statutory regulations is set out in the schedule below. They reflect and influence the broad common law duty to restrict exposure to asbestos (and other dangerous substances) of an employee so as not to create a foreseeable risk of injury.

12. See also *Bryce* v *Swan Hunter Group plc* [1987] 2 Lloyd's Rep 426.

Regulation	Citation and Current Status
Asbestos Industry Regulations 1931	Repealed
Factories Act 1937	s.4, s.47: Repealed:
Occupiers' Liability Act 1957	In force, but see *Fairchild* v *Glenhaven Funeral Services* [2002]: Court of Appeal, below.
National Insurance (Industrial Injuries) (Prescribed Diseases) Regulations 1948	SI 1948/1371: Repealed
Factories Act 1961	s.4, s.63 Repealed
The Asbestos Regulations 1969	Reg 3(2): Repealed
The Asbestos Products (Safety) Regulations 1985	SI 1985/2042:
Asbestos (Prohibitions) Regulations 1985	SI 1985/910
Control of Asbestos at Work Regulations 1987	SI 1987/2115: relevant regulations 4, 5, 6, 7, 8, 9, 10, 17, 21: Repealed: see 2002 Regulations
Asbestos Products (Safety) (Amendment) Regulations 1987	SI 1987/1979: repealed
Asbestos (Prohibitions) (Amendment) Regulations 1988	SI 1988/711
Control of Asbestos at Work (Amendment) Regulations 1992	SI 1992/3068: Revoked by the Control of Asbestos at Work Regulations [2002]
The Control of Asbestos in the Air Regulations 1990	SI 1990/556
The Asbestos (Prohibitions) Regulations 1992	SI 1992/3067
The Asbestos (Prohibitions) (Amendment) Regulations 1999	SI 1999/2373
The Asbestos (Prohibitions) (Amendment) No. 2 Regulations 1999	SI 1999/2977 Implementing EC Directive: 99/77/EC: see also EC Directive: 76/769/EEC:
Control of Asbestos at Work (Amendment) Regulations 1998	SI 1998/3235: repealed by Control of Asbestos at Work Regulations 2002

The Control of Asbestos at Work Regulations 2002

These regulations came into force on 31st October 2002. The main regulations are:

Regulation 3: imposes a statutory duty on an employer in respect of the employees in relation to "non domestic" premises.

Regulation 4: imposes a duty on the "duty holder"[13] to manage asbestos in non-domestic premises. This includes an assessment of whether asbestos is present on the premises and if so, what is the level of risk posed by the asbestos. Measures then have to be put in place to manage the risk.

Regulation 6: precludes an employer from carrying out work that exposes employees to asbestos, unless there is a suitable and sufficient risk assessment. The prescribed contents of the risk assessment are set out in the regulations.

Regulation 9: employers are required to provide adequate information, instruction and training to employees so that they are aware of the risks of exposure to asbestos. This duty is a continuing one and has to take account of adaptations to the system of work.

Regulations 10, 11 and 12: so far as reasonably practicable employers have to prevent the employees from being exposed to asbestos. Where it is not practicable to prevent exposure to asbestos then specific control measures have to be set in place to avoid or minimise the release of asbestos.

Regulation 13: Employers have to provide suitable protective clothing for the employees.

Regulation 15: Every employer is under a duty to prevent (or where this is not practicable) to reduce to the lowest reasonable level the spread of asbestos from any place where work under the employer's control is carried out.

---

13. Defined in the regulation as a person who by contract or tenancy (or who has control of the premises) has an obligation to repair or maintain the non-domestic premises. There can be more than one "duty-holder". As to the determination of liability issues in asbestos related claims under employer's liability policies: see *T and N Ltd (in administration)* v *Royal and Sun Alliance Plc* [2003] EWHC 1016 (Ch).

Regulation 16 and 17: impose a duty to keep the employer's premises clean and to have designated areas for this type of work (including air monitoring: reg 18).

Regulation 28: now provides a defence of due diligence arising out of breach of these regulations. This has implications for a civil claim. The contention that there is an absolute standard of compliance required by the employer and the standard of duty of care have to be considered in the context of whether the employer has taken all reasonable steps to make and keep safe the employees.

The Occupiers' Liability Act 1957

The Occupiers' Liability Act 1957 imposes a duty on the "occupiers" of premises but care has to be taken as to how far that statutory duty is applicable to asbestos exposure. It depends on who is exposed to the asbestos and in what circumstances. For example, at common law there is no duty on an occupier of premises in relation to independent contractors working at the premises, when the danger of injury arises from the nature of the work and not the nature of the premises. In *Fairchild* v *Glenhaven Funeral Services Ltd (t/a GH Dovener & Son)*,[14] a widow of a former employee of the defendant who died from malignant mesothelioma due to asbestos exposure appealed against the dismissal of her claim. It was held that where an occupier under the 1957 Act had engaged an independent contractor to work at their premises that resulted in asbestos exposure, the occupier had no liability under the Occupiers' Liability Act 1957, s.2(2). The section related to liability that would arise from the condition of the premises and not the activities carried on at the premises (i.e.working with asbestos).

Foreseeability

The case law

Having established the date of knowledge of the risk of injury of an employer, then a claimant has to prove that injury was a reasonably foreseeable consequence of that breach of duty. In practice, this means proving a failure to take reasonable measures

14. [2002] 1 WLR 1052.

to prevent the claimant being exposed to asbestos. Case law is fact sensitive, but the following cases provide some guidance:

(a) *Edgson* v *Vickers Plc*:[15] it was held that dry sweeping was a negligent process that was covered by the Asbestos Regulations 1969 Regulations 3(2) and 15(1).

(b) *Bryce* v *Swan Hunter Group Limited*:[16] concerned the death of a ship painter from mesothelioma. It was held that the employer was under a duty to reduce the limit of exposure to asbestos and that its failure to take such reasonable action as was required was a breach of its duty of care.

(c) *Shell Tankers (UK) Limited* v *Jeromson and Cherry Tree Co Limited*;[17]

- the provisions of the Asbestos Industry Regulations 1931 did not apply only to "the asbestos industry". They applied to factories and workshops where asbestos was used, not merely where asbestos was manufactured.[18]
- There was an absolute requirement to provide an exhaust. The defendant employer's liability was not limited by "reasonable practicability".
- A reasonably prudent employer would take precautions as to the exposure to asbestos, or at least would make the appropriate enquiries about relevant precautions that were required.

In *O'Toole (Deceased)* v *Iarnod Eireann Irish Rail*,[19] the claimant was a railway carriage builder between 1950 and 1955. He developed mesothelioma consequent on exposure to asbestos at work and died in 1987. He sprayed carriages with asbestos and the process created a significant amount of dust which was controlled by his employer in addition to a contractor. In finding for the claimant it was held that:

- the claimant had to prove that the dust made in the industrial process created a foreseeable risk of injury, but

15. [1994] ICR 510.
16. [1987] 2 Lloyd's Rep 426.
17. [2001] EWCA Civ 101.
18. *Watt* v *Fairchild Shipbuilding and Engineering Co Ltd* [1999] SLT 1084 was not followed.
19. [1999] QBD 19.2.99 Unreported.

the particular type or extent of the respiratory injury did not have to be proved.

- The employer was in breach of its duty to reduce to a minimum the exposure to the risk of pulmonary injury.

Causation

A claimant has to prove that the breach of duty of the defendant has caused the injury in question. This issue is covered in detail in chapter 2 on causation but in summary the following points are to be noted:

(1) In cases of "divisible" respiratory disease (including asbestosis, emphysema and bronchitis arising from exposure to dust) there will be an apportionment of causation. In contrast in the case of "non-divisible" injuries (e.g. mesothelioma) there is no apportionment between competing causes.

(2) A defendant is liable only to the extent that the negligence has contributed to the "divisible" injury.

(3) Damages for a "divisible" injury will be reduced to reflect other contributory factors and examples include:

- Pre-existing conditions that are exacerbated by the negligence.
- Smoking and other activities that adversely affect respiration.
- Any injury that is established prior to when the duty of care was imposed.
- Non-tortious exposure, (for example to coal dust).

(4) The case law set out below provides some guidance on how the assessment of quantum is made in the face of relevant causal factors.[20]

20. In the case of coal dust see *Griffiths* v *British Coal* [1998] Lawtel 23.1.98.

Practice Points: A Claimant's perspective

(1) It is vitally important to obtain full instructions from a claimant at the earliest opportunity. Where the death of the claimant is imminent then a full signed statement ought to be taken. At this early stage there will probably be limited disclosure but nevertheless a claimant (including his family) has to set out a full work history to demonstrate (a) the periods of exposure and (b) the nature of the exposure at the workplace.[21] Full particulars of the system of work and the risks of injury created are essential. These can be obtained from both the claimant and his family, and also any relevant work colleagues.

(2) Further documentation[22] will include medical records plus any relevant documentation from the D.S.S. Benefits Agency. These have to be closely considered for the following reasons:

(a) to ascertain when the claimant first presented with symptoms to his general practitioner, as this is relevant to limitation. It may also reveal whether the claimant attributed them to the work conditions;

(b) with an application for state benefits the claimant will probably have been medically examined and he will invariably have given an explanation of the circumstances of the medical condition and its suspected causes;[23]

(c) identify any potential inconsistencies between the present instructions and what the claimant has stated on previous occasions as identified above.

(3) On full disclosure, detailed consideration of all the relevant documentation relating to the system of work has to be made. This enables important factors to be clarified including;

- an assessment of the knowledge of the defendant of the risk of injury;

- whether limitation will be an issue;

21. This information will enable relevant insurers to be tracked down and their status to be assessed.
22. Assuming there has been no post mortem/inquest at this stage.
23. For example, Industrial Injuries Disablement Benefit.

- the nature and extent of the risk of injury posed, for example, by the exposure to asbestos.

(4) In turn, this should also provide the evidential basis for the specific allegations of breach of duty by the defendant. This can then be further investigated by a health and safety expert.

Investigations may reveal that a claimant worked for a series of existing companies each of which is potentially culpable. Alternatively, an alleged defendant company may have been dissolved. Steps will have to be taken to identify the relevant companies (and insurers) and if necessary to restore them to the companies' register, pursuant to s.651 of the Companies Act 1985. Judgment can then be obtained. The cause of action will have accrued at the date of knowledge of the injury and this is not affected by a subsequent dissolution.[24]

(5) Expert evidence is required at an early stage and this will include:

(a) Medical evidence on causation. In cases of mesothelioma and asbestosis unless the medical evidence is supportive on causation then the claim will fail. Assuming that condition is satisfied, then further medical evidence is required on;

- life expectancy; and
- nursing care; and
- percentage apportionment between constitutional factors and the asbestosis/mesothelioma.

(b) An expert in health and safety is required. Close analysis is required of all the relevant documentation on the system of work and health and safety at the relevant times.

(c) Analysis has to be made of the potential issues of limitation and provisional damages issues (including the evidence in relation to each) at an early stage.

24. *Smith* v *White Knight Laundry Limited* [2001] EWCA Civ 660. [2002] 1 WLR 616. See generally chapters 4 and 5 on limitation. See also Third Party (Rights Against Insurers) Act 1930.

(d) A detailed signed statement from the claimant is needed setting out all the relevant facts and matters on which expert opinion will be based.[25]

Practice Points: A Defendant's Perspective

It is necessary to determine as soon as possible the extent to which liability is in issue. This will include an assessment of the following:

(1) Particulars are required from the claimant's solicitors on:

- the circumstances of the alleged exposure;
- the nature and extent of the exposure;
- the period of employment and whether this is relevant to the insurer's cover.

Effectively, a request ought to be made for a detailed statement or statement of case. This will facilitate investigations but also clarify the claimant's case at an early stage.

(2) Assess whether there is an issue on the knowledge of the defendant of the dangers of asbestos, enabling liability to be contested on the basis that the disease was contracted prior to the date of knowledge. In practice this is unlikely to be the case.

(3) Specific investigations may be required as to the relevant defendant (in terms of insurance cover over that period) and/or whether there was more than one company against which proceedings ought to be issued. Part 20 proceedings may be a necessary step.

(4) Whether the defendant company operated a system of work that reduced the exposure to asbestos to a safe level. Relevant documentation will include minutes of company meetings, health and safety assessments and all other documentation on the system of work.

25. It also provides consistency in the claimant's version of events.

(5) This needs an analysis of the relevant documentary and medical evidence. For example, medical records may potentially highlight when the claimant first had knowledge of his injury, or contradict his version of events.

(6) Expert evidence is important and this includes:

- Expert evidence from a heath and safety expert at an early stage, on breach of duty, based on all the relevant documentation. What precautions were taken and what precautions should have been taken under the regulations.
- Medical evidence on causation and life expectancy.
- Medical evidence on provisional damages, including the percentage chances of a "serious deterioration" in the claimant's future condition or, disease.[26]

Limitation

Specific Points

Limitation is covered in detail in chapters 4 and 5 and so this section is restricted to specific points on asbestosis. In asbestos related cases the knowledge of injury of the claimant is vital in establishing when the cause of action accrued pursuant to sections 11 and 14 of the Limitation Act 1980.[27] In many cases, a claimant will have contracted a serious respiratory disease many years before the presentation of significant symptoms. Moreover, such cases are difficult to investigate, particularly as many claimants smoked or suffered from constitutional conditions that can blur the date of knowledge of injury. These factors tend to weigh heavily against knowledge of a significant injury by the claimant being found at or about the time of the original employment. Even if the claim is time barred on say, knowledge of a mild respiratory disorder, this may not necessarily preclude a claim based on more serious subsequent pathology, such as asbestosis or mesothelioma.[28]

26. See chapter 3 on this issue.
27. Note that in cases of asbestosis and mesothelioma the statutory regulation prior to the Limitation Act 1980 can be relevant.
28. There are discrete limitation periods for discrete medical conditions. See Chapters 4 and 5.

Repealed Limitation Acts

In this type of case the limitation issues may extend back before the commencement of the Limitation Act 1980 and so the provisions of the previous limitation statutes need to be analysed, including the Limitation Acts of 1939, 1963 and 1975. As to the operation of the Limitation Act 1963, the following cases are relevant on the criteria for disapplying the limitation period:[29]

- *Smith* v *Central Asbestos Co* [1971] 3 W.L.R. 206.
- *Drinkwater* v *Joseph Lucas (Electrical)* [1970] 3 All E.R. 769. Incorrect advice from trade union could justify extension of the limitation period.
- *Newton* v *Cammell Laird & Co (Shipbuilders and Engineers)* [1969] 1 All E.R. 708.
- *Arnold* v *Central Electricity Generating Board* [1988] AC 228: relating to a claim subject to the Limitation Act 1939.

Quantum of Damages

To provide some guidance on quantum a series of cases is set out in the schedule below, which highlight the most recent awards of general damages. Consideration also needs to be given to:

(a) the relevant apportionment between constitutional factors and symptoms caused by the defendant's negligence, with the medical evidence covering the issue;

(b) nursing care;

(c) provisional damages namely the right to return to court if a claimant's condition or disease deteriorates significantly which could not be calculated at trial. This is more likely where the future risks are not known;[30]

29. See *Griffiths* v *British Coal* [1998] Lawtel 23.1.98: which held there was no liability for claims in miners' respiratory diseases before 1.1.48: see also *Arnold* v *Central Electricity Generating Board* [1988] 1 AC 228.
30. *Bahardo* v *Mayer Newman and Co Limited* [2002] Lawtel 21.1.02: the case also provides a helpful assessment of quantum in a claimant whose disability was increased by non-asbestos related factors.

(d) a dependency claim[31] and specifically, where relevant, multipliers are calculated from date of death in relation to dependency claims;[32]

(e) the terms of any settlement reached between the parties;[33]

(f) Statutory compensation schemes.[34]

31. See *O'Loughlin* v *Cape Distribution Limited* [2001] PIQR Q8.

32. Multipliers are to be calculated from the date of death: *White* v *ESAB Group (UK) Limited* [2002] PIQR Q6: As to the correct approach in a "lost years" claim see *Phipps* v *Brooks Dry Cleaning Services Ltd* [1996] Q100.

33. *Jameson* v *Central Electricity Generating Board (No.1)* [2000] 1 A.C. 455: *Green* v *Vickers Defence Systems Limited et al* [2002] *Times Law Reports* 1.7.02.

34. See British Coal Respiratory Disease Litigation: *Griffiths* v *British Coal* [1998] *supra*.

Name of Case	Award and Comment
Blair v *Michelin Tyre Plc* [2002] MLC 666	C male, exposed to asbestos between the ages 23 and 51. At trial (aged 68) claimed damages for an asbestos related disease. Liability was admitted. C also suffered from chronic obstructive pulmonary disease attributable to smoking. C's respiratory condition meant he was 40% disabled. 70% of that disability was due to asbestos exposure. *General damages £40,000.* Further 225 hours was justifiable for a gratuitous care claim: An award of £20,300 for future care.
McKenzie v *Barclay Curle Limited* [2002] S.L.T. 649;	Claimant, 61, a former shipyard engineer, claimed damages for pleural plaques/thickening and asbestosis caused by exposure to asbestos at work. At risk of mesothelioma or cancer. Held: C suffered from a 50% disability, half of which was attributable to asbestos related conditions. *General Damages: £35,000*
Bahardo v *Mayer Newman & Co Ltd* [2002] 3 Q.R. 5	Male, aged 61, exposed to asbestos dust between 1969 and 1983. Diagnosed with asymptomatic pleural plaques and pleural thickening. Overall respiratory disability was 30 to 40%. 25% was caused by smoking. Respiratory disease caused by asbestos exposure was 22.5 to 30%. C would have retired at 60 in any event. Hobbies severely affected and could not walk long distances or climb stairs easily. 20% risk of further pleural thickening, and 15–20% chance of cancer from asbestos exposure. 8–15% chance of mesothelioma with a reduction of life expectancy by 3 years from unrelated disease and 2–4 years from asbestos exposure. *General Damages: £20,625 discounted by 10% on a provisional damages basis including £10,000 for loss of DIY ability.*

Name of Case	Award and Comment
Hunt v *A Robinson & Co* [2001] 3 Q.R. 18	C, male 59 exposed to asbestos dust between 1966–1972. Diagnosed with pleural mesothelioma but worked up to four months before his death. Then suffered breathlessness and severe chest pain. Life expectancy was reduced from 18 years to 7 years, due to his past history of myocardial infarction and blood pressure. C would probably have retired at the age of 64 in any event. *General Damages: £50,000*
Cook v *Wyvern Structures Ltd (In Liquidation)* [2001] S.L.T. 1212	C, a plumber's mate in a shipyard. Died in 1992 from a respiratory disease, probably asbestosis. Exposed to asbestos in more than one employment from 1942 to 1965. Causation could not be proved on the basis of the medical evidence. *Damages would have been £10,000 based on 35% disablement for six years. £40,000 for death; £2,500 for services provided by his spouse.*
Wilkins v *William Press Ltd* [2001] 3 Q.R. 19	C, male, 40 suffered mesothelioma following exposure to asbestos at work. Experienced increasing pain in his chest, shoulder and back and underwent decortication of his left lung. The pain was controlled by drugs and he underwent radiotherapy and chemotherapy and died in severe distress. *General damages: £52,500.*

Name of Case	**Award and Comment**
White v *Port of London Authority* [2001] 3 Q.R. 19	C male, aged 81 at death, contracted malignant mesothelioma due to asbestos exposure as a dock worker between 1949–1967. Symptoms included breathlessness with pleural effusion. He also underwent partially successful surgery that left him with a pump and a tube in his chest. C died 9 months after the presentation of his symptoms. Life expectancy was reduced by one year in any event: *General Damages £40,000. Past care award (assessed at 16 hours per day for the last seven weeks of his life): £3,920*
Hunter v *British Railways Board* [2001] Current Law 1714	C male, aged 75 at death. Previously exposed to asbestos between 1939 and 1986 at work. Diagnosed with mesothelioma caused by asbestos exposure. C was short of breath and was in significant pain. He developed a pleural effusion. Prior to mesothelioma, he had been a youthful and active man. *General Damages: £50,000.*
Kelly (Deceased) v *Smith & Partners Ltd* [2001] 3 Q.R. 19	C male, 70 at death, suffered malignant mesothelioma due to exposure to asbestos at work between 1963–1966. He suffered increasing breathlessness up to his death in 1998 (16 months total symptoms from diagnosis). He underwent radiotherapy and required intensive nursing for the last 3 months of his life. *General Damages: £47,500.*

Name of Case	Award and Comment
Holmes-More v *Tenon Ltd* [2001] 3 Q.R. 18	C, male, 49 exposed to asbestos at work from 1968–1975. Diagnosed in 1997. By 1999 there was no further deterioration and C was re-diagnosed with pleural thickening and pleural plaques. His condition was 20% due to exposure to asbestos at work. C was still able to work. His life expectancy was reduced by one–two years by the risks of malignant disease. There was a 30 to 50% risk of further pleural thickening and 5% risk of lung cancer. Permission was given to apply for further damages if C developed either asbestosis, mesothelioma or asbestos related lung cancer. *General Damages: £25,000 (provisional award). Loss of domestic capacity award: £13,500.*
Meachen v *Harwich Dock Co Ltd* [2000] 5 Q.R. 7	C, male, aged 66 exposed to asbestos dust between 1973–1983 working as a docker. He developed pleural plaques and pleural thickening, diagnosed in 1995. C had a total respiratory disability of 60%, with 10% due to asbestos related disease and 50% due to emphysema. 1% risk of asbestosis, a 3% risk of mesothelioma, a 15% risk of lung cancer. No evidence of anxiety. *General damages (provisional award): £10,000.*
O'Toole (Deceased) v *Iarnod Eireann Irish Rail (Quantum)* [1999] 99(6) Q.R. 5	C, male, was employed as a railway carriage builder with D between 1950–1955. C developed breathlessness due to mesothelioma diagnosed in 1987. C underwent surgery for a pleural effusion in addition to radiotherapy. He died in November 1987. It was held that D was 30% to blame for C's mesothelioma and D2, (an additional defendant) 70%: *General Damages £40,000.*

Name of Case	Award and Comment
Jeffrey v *Cape Insulation Ltd* [1999] 99(2) Q.R. 5	C, male, 67, was exposed to asbestos dust when he was 36, for a period of about eight months. 25 years later C suffered chest pain and partial loss of breath. 31 years after exposure he was diagnosed as suffering from asbestosis. C was assessed as 50% disabled, consisting of 30% attributable to asbestos and 20% attributable to chronic bronchitis and emphysema. C was limited in his leisure activities and required three hours care per week. His condition would deteriorate with a risk of mesothelioma and lung cancer. Loss of life expectancy of eight years, mostly due to the asbestos exposure. *General Damages: £38,000. Nursing care award: £13,000.*
Worrall v *Powergen Plc* [1999] P.I.Q.R. Q103	C male, 63 at death, suffered from mesothelioma, due to asbestos exposure in 1957–1970s. C suffered symptoms from 1995. Chemotherapy was unsuccessful. C also suffered a tumour through the chest wall with a large portion being visible outside the body for several months before he died in 1997. *General Damages (under the 1934 act): £45,000. Award for past family care (after deduction of Disability Living Allowance): £8,160.*
Horne v *Prescot (No. 1) Ltd* [1999] 99(2) Q.R. 6	C, male 64 at trial, was exposed to asbestos for about 4 years between the ages of 16 to 23. At 60 pleural plaques were diagnosed. C suffered shortness of breath on exertion. An overall risk of developing lung disease, including asbestosis and mesothelioma or lung cancer, at 16%. 10% risk of developing mesothelioma, and 1% risk of lung cancer. C's life expectancy was reduced by one year as a result of the exposure to asbestos. The overall risk of any deterioration or development of other asbestos related disease was very low. *General Damages: £12,500.*

Name of Case	**Award and Comment**
Lightbody v *Upper Clyde Shipbuilders Ltd* [1998] S.L.T. 884;	C, a retired shipbuilder, 59, contracted asbestosis due to exposure to asbestos at work. C had suffered from heart problems and even underwent surgery unrelated to his work. Moderate breathlessness for a number of years was attributed to asbestosis. The asbestosis was minor to moderate with a 2% chance of developing mesothelioma. *General damages: £15,000.*
Farrelly v *Yarrow Shipbuilders* [1994] S.L.T. 1349	Electrician, 62 contracted mesothelioma from exposure to asbestos. Domestic services based on a multiplier of seven years. Services given to C by his wife assessed at £50 a week for the ensuing nine months and then £385 per week for the final 12 weeks of his expected life. *General Damages: £45,000*
Margereson v *J.W. Roberts Limited* [1996] PIQR 154	Male; mesothelioma contracted from inhaling asbestos in the atmosphere from the defendant's factory. *General Damages*: *£50,000 and £65,000.*
Myles v *Glasgow DC* [1994] S.C.L.R. 1112	C, 55, suffered asbestosis from employment between 1968 to 1990. Extent of causation disputed as C was obese and smoked. Leisure activities were severely curtailed and he retired early. Provisionally assessed at 15% disabled. 2% risk of mesothelioma and 5% risk of lung cancer. *General Damages £27,000*: C was 50% disabled as a result of breathlessness but 10% attributable to obesity and 15% to asthma. Consequently, 25% was attributed to his exposure to asbestos dust.

Name of Case	Award and Comment
British Coal Respiratory Disease Litigation: *Griffith* v *British Coal* [1998] QBD: Lawtel 23.1.98	A government scheme to compensate 65,000 miners for respiratory problems arising from dust inhalation. Damages were apportioned to take account of smoking but awards were up to £50,000: £3,000–12,000 was awarded to widows of those miners who died after 20 years' employment.
Nicol v *Scottish Power plc* [1998] S.L.T. 822;	C male, 52, scaffolder exposed to asbestos dust for 20 years at work. In 1994 diagnosed with bilateral pleural plaques. C's symptoms were caused by smoking, but the risks of developing mesothelioma and cancer had been increased. *General Damages £13,500.*

Overall Summary

The schedule sets out a summary of the main points of the chapter in addition to providing practical guidance for defendant and claimant practitioners on the handling of asbestos related claims.

Summary Point	Claimant's Perspective	Defendant's Perspective
Duty of care • Owed at common law and under statutory regulation. Foreseeability at common law is subject to knowledge of risk of injury by employer. No such qualification on statutory regulation unless specifically provided for. • Statutory regulations mainly determine the nature and extent of the duty of care.	• Identify the correct employer at each stage of exposure and clarify the relevant insurer. • Ensure that correct statutory regulations apply to relevant period of exposure. • Obtain a signed statement from the claimant detailing all exposure periods. Same action can be taken in relation to colleagues. • Obtain early disclosure if possible.	• Clarify relevant employers and who is the relevant insurer. • Identify nature and periods of exposure by particulars from claimant. • Tie claimant down early to version of events.
Foreseeability Proof of a foreseeable risk of injury. Knowledge of injury by defendant probably from 1932 onwards: Asbestos Industry Regulations 1931	• Reliance to be placed on HSE and other publications to show defendant's knowledge of risk of injury.	• Defendants will probably be unsuccessful in claiming no knowledge of the risk of injury.
Breach of statutory duty: Standard of the competent employer: (*Stokes* v *Guest et al*). Breach will depend on the statutory regulations at the relevant time: note new Asbestos 2002 Regulations: "due diligence" defence to criminal offences.	• A detailed statement is required from claimant on system of work and extent of exposure. • Expert evidence on system of work and breach of regulations is required. Ensure that expert is properly qualified in this specialist area. Breach of duty and causation has to be covered in the report.	• Same as claimant: ensure statements are obtained from witnesses, including line managers and colleagues of claimant if possible, on issue of system of work. • Obtain expert evidence. Breach of duty and causation has to be covered in the report, based on relevant statutory regulations.

Summary Point	Claimant's Perspective	Defendant's Perspective
Limitation Limitation is invariably an issue. With knowledge of injury (s.14) outside time limits then an application under s.33 is required.	• Clarify with claimant when connection first made between symptoms and work. Ensure these particulars are contained in a statement. • Is this a s.14 issue on knowledge or is an application under s.33 needed? An application at an interim stage or at trial? • Obtain any DSS/Benefits Agency records and assess version of claimant for consistency.	• Assess when claimant first knew of these symptoms and that they were connected to work. • Obtain DSS/Benefits Agency records and see what version claimant has given in these. • Medical expert to comment. Ensure that limitation is pleaded fully. • Consider if this is a type of injury such that only part of the claim may be statute barred.
Causation • Causation has to be assessed on basis of *Fairchild* v *Glenhaven* and *Holtby* v *Brigham*: a material contribution to the injury is required. • Note the differing approach to causation in mesothelioma and asbestosis cases, based on "divisible" and "indivisible" injuries.	• Obtain medical evidence on issue of causation and contribution to the injury. • Ensure the medical evidence states on balance that there was a material contribution to the injury. • Assess whether the injury is "divisible" on causation or not and have this issue covered in medical evidence.	• Analyse whether other defendants have made a contribution. • Assess on medical evidence the extent of contribution. • Ensure medical expert covers this issue in detail.

Summary Point	Claimant's Perspective	Defendant's Perspective
Quantum • With "divisible" asbestos injuries (asbestosis) there will be percentage apportionment with non-tortious factors (e.g. smoking). • Provisional damages and future care regimes have to be considered carefully. Life expectancy needs to be clarified.	• Medical evidence to assess the non-tortious contributions and extent to which this would have affected present condition in any event. • A costed care regime is required and this should take account of non-tortious symptoms and life expectancy. • Provisional damages must be covered in medical evidence.	• Obtain medical evidence on contribution and whether claimant would have been substantially in this condition in any event. Percentage attribution is required. • Obtain evidence on care, and life expectancy, taking account of the above. • Clarify provisional damages with expert. Avoid provisional damages if possible.

Section 2: The Medical Perspective

Overview

The decline in traditional heavy industry in the UK during the last 40 years, particularly coal mining, together with stricter health and safety legislation, has resulted in a reduced incidence of respiratory diseases due to inhaled inorganic dusts, such as coal worker's pneumoconiosis, asbestosis and silicosis.

By contrast, there has been an increased incidence of cases of occupational asthma, which is now the commonest cause of occupational respiratory disease. However, due to the long latent period of over 40 years, deaths from mesothelioma due to previous asbestos exposure remain common, and may not peak until 2020.

Basic Anatomy

The upper airway is comprised of the mouth, nose and throat (pharynx). The mucus-secreting tissues of the nose lie in folds (turbinates) lined with hairs.

The lower airway commences at the trachea, which branches in the neck to supply air to each lung via the left main bronchus and the right main bronchus. The bronchi continue to divide, and the smallest branches are known as bronchioles (figure 1).

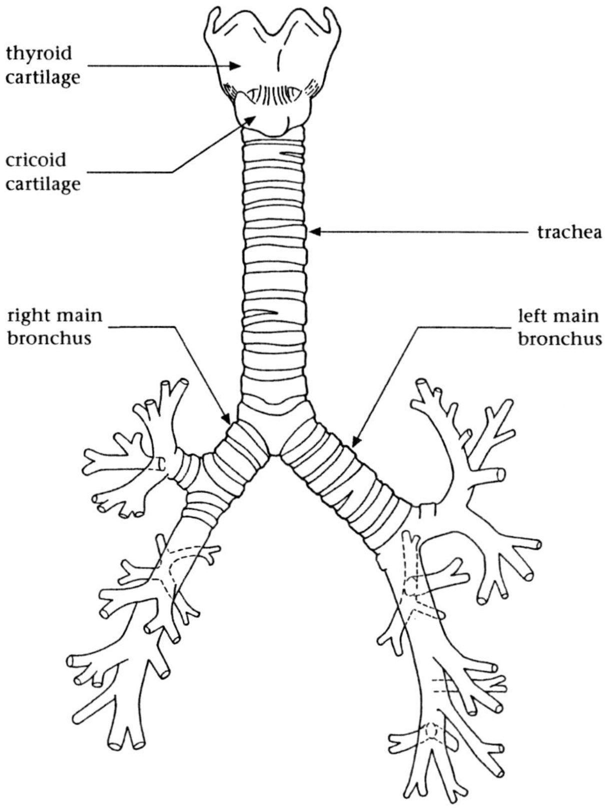

Figure 1

They have a wall containing smooth muscle, which is capable of involuntary contraction in response to a number of stimuli. They terminate in structures known as the alveoli, where gas exchange takes place with the pulmonary circulation (figure 2). The lung tissues, which provide an elastic supporting structure for the airways and related blood vessels, are known as the lung parenchyma.

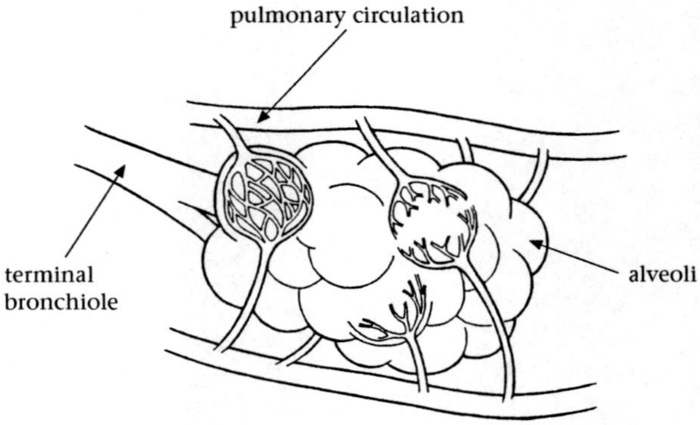

Figure 2

The pleural membranes line the inner surface of the chest wall, the upper surface of the diaphragm, the surfaces of the pericardium and mediastinum (the parietal pleura) and the outer surfaces of the lungs (visceral pleura).

The potential space between the membranes contains a tiny amount of pleural fluid, which provides hydrostatic pressure. This allows the visceral pleura to remain closely adherent to the parietal pleura during normal respiration.

Normal Clearance of Dust Particles

The normal dust clearance system should work as follows. The nasal turbinates provide a filter for most particles over 15µm in diameter. They are carried to the pharynx in mucus and swallowed.

The trachea, bronchi and bronchioles are lined by mucus secreting cells bearing tiny filaments (cilia), the mucociliary epithelium. This lining provides a further filter for most particles down to about 5µm, and some of the smaller particles. They are carried by the mucus and cilia back to the pharynx and swallowed.

The presence of dust particles on the mucociliary epithelium stimulates mucus production, which assists clearance. If the particles are also irritant, bronchial nerves are stimulated, resulting in cough, contraction of the muscles with consequent airway narrowing, and further outpouring of secretions.

Small particles which get beyond the mucociliary epithelium will reach the alveoli. They become coated in the fluid lining the alveoli ("surfectant"), and are then engulfed (the process of "phagocytosis") by white cells migrating from the adjacent tissues (macrophages). The macrophages then ferry the particles back to the bronchioles, where they are carried by the mucociliary system to the pharynx and swallowed.

Basic pathophysiology

Chronic bronchitis

Prolonged exposure to lung irritants may lead to a state of permanent excess mucus production and obstruction of airflow, a condition known as chronic bronchitis. The commonest cause in the population as a whole is cigarette smoking. This is partly as a result of excess growth of the mucus producing cells in the trachea and bronchi.

Emphysema

There is often associated loss of the lung parenchyma (emphysema), probably as a result of the toxic effects of smoke on the macrophages. This leads to collapse of airways through loss of elastic recoil, trapping of air, and loss of the blood supply needed to maintain effective exchange of gases.

Chronic bronchitis and emphysema result in fixed reduction of the ventilatory capacity of the lungs, predominantly due to limitation of expiratory airflow. This condition is known as chronic obstructive pulmonary disease ("COPD").

Principal effects of dust

In the occupational setting, inorganic dusts do not usually cause significant bronchial inflammation. Their disease effects are mainly produced through damage to the lung parenchyma:[35]

(a) Overload of the normal dust clearance systems, leads to an accumulation of dust particles in the alveolar macrophages, which eventually leads to their death.[36]

(b) When the macrophages become overloaded and die, the surrounding tissues also become inflamed. The affected tissues either die or they are replaced by scar tissue (fibrosis).

(c) This process of inflammation and fibrosis eventually interferes with normal ventilation by making the lungs inelastic and immobile.

(d) The rate at which this process occurs depends on how toxic the dust is to the macrophages.[37]

The pneumoconioses

Pneumoconiosis is the generic term for diseases caused by inhaled dust lodging in the lungs, irrespective of their effects, but excluding asthma and cancer. Asbestosis is one example. Other examples due to inorganic mineral dusts[38] include:

(a) coal worker's pneumoconiosis (coal dust)

(b) silicosis (silicon dixoide)

(c) berylliosis (beryllium)

(d) talcosis (magnesium silicate)

(e) siderosis (iron oxides)

---

35. An exception is COPD related to coal dust exposure, which after much controversy as to the confounding effects of smoking, has recently become a prescribed industrial disease (PD D12), provided there is evidence of underground exposure for at least 20 years, and a sufficient degree of obstruction.
36. Long, thin particles (fibres) are more dangerous, since they are far more difficult to clear.
37. Crystalline forms of silicone dioxide, such as quartz, are far more toxic to macrophages than coal dust and asbestos fibres.
38. Organic dusts such as cotton and flax have different effects ("byssinoisis").

Asbestos Varieties

The most important group of fibrous mineral dusts is the asbestos family, a group of compound metallic silicates, which crystallize as very long fibres. Chrysotile has a different structure from the other forms of asbestos, including crocidolite (blue asbestos) and amosite (brown asbestos). The latter are known as "amphibole" fibres. The white, blue and brown varieties are all used commercially.

White asbestos

White asbestos[39] has long, soft fibres, which can be weaved into heat-resistant textiles, or mixed with cements or plastics to form lagging for pipes, light, rigid construction materials, and air filters.

Blue asbestos

Blue asbestos has greater resistance to chemicals such as acids, however, the fibres are shorter and stiffer than those of white asbestos, and difficult to weave.

Brown asbestos

Brown asbestos is used for insulation and as a reinforcing material for certain plastics. It is also difficult to weave, or to use mixed with cements.

Asbestos fibres are also known inducers of cancer (carcinogens). Exposure is associated with cancer of the bronchus (bronchial carcinoma), and cancers (mesothelioma) of the pleural membranes. Mesothelioma sometimes also affects the peritoneum, the membrane covering the abdominal organs, and the pericardium, the sac which surrounds the heart.

39. There has recently been considerable debate about the safety of white asbestos, which is present in many buildings in the form of asbestos board. Chrysotile fibres are thought to create less risk of asbestos-related diseases than amphibole fibres, as they are more readily cleared from the lungs. However, the removal of asbestos products from buildings creates its own risks.

Asbestos-Related Diseases

Asbestosis

Pneumoconiosis due to asbestos dust is known as asbestosis. It is a gradually progressive condition. There are no characteristic symptoms. In the early stages there may be no symptoms at all. As the disease progresses there is a gradual onset of breathlessness and cough. In smokers, this may be attributed initially to this habit.

Examination in the later stages may reveal bulbous swelling of the nail pulps (clubbing), and crackles in the lungs, especially at the bases. Investigations which are commonly undertaken include:

(a) *Chest x-ray*. Fibrosis may appear as linear and irregular shadows, which gradually progress to a "honeycomb" appearance.

(b) *Lung function tests*. These show a characteristic "restrictive" pattern due to reduced lung compliance. There is reduction of the total lung capacity for air (TLC), and the volume of air that can be forcibly exhaled (FVC), without significant obstruction of airflow. There is also reduction in the ability of the lungs to exchange gases, the "transfer factor" (TLCO).

(c) *CT (computerised tomography) or MRI (magnetic resonance imaging) scanning of the chest*. They are useful in picking up early cases, and differentiating pleural from parenchymal disease.

There is no specific treatment. Cessation of exposure to asbestos dust and smoking is essential. Sometimes oral steroids are employed, but only a small proportion of patients will respond. In the worst cases, with end stage respiratory failure, heart and lung transplantation is sometimes attempted. However, the major cause of death is from primary lung cancer or mesothelioma.

Asbestosis is a prescribed industrial disease (PD D1).

Non-malignant pleural diseases

Asbestos exposure results in a number of non-malignant conditions affecting the pleural membranes.

(a) *Pleural plaques* are benign circumscribed areas of pleural thickening due to fibrosis, which usually occur on the surface of the parietal pleura. The visceral pleura is not usually involved. The characteristic chest x-ray appearances, including calcification, take many years to develop. However, the plaques are readily detected by CT or MRI scanning of the chest in the earlier stages. Pleural plaques do not give rise to symptoms, but are important as an indicator of previous asbestos exposure. They do not become malignant. No treatment is required. They are not a prescribed industrial disease.

(b) *Benign pleural effusions* are a collection of fluid in the potential space between the parietal and visceral pleura due to inflammation of the pleura (pleurisy). A pleural effusion is often transient and there may be no symptoms. Sometimes there is a flu-like illness, breathlessness, and pain in the affected side of the chest. The presence of an effusion will be revealed by examination and chest x-ray. However, further investigations will be directed to excluding more serious conditions such as malignant mesothelioma (see below). Steroid therapy is sometimes employed to hasten the resolution of pleural effusions. They often resolve completely, leaving little residual pleural thickening. Benign pleural effusions may sometimes progress to diffuse pleural thickening. However, the condition is not a prescribed industrial disease.

(c) *Diffuse pleural thickening* is a condition of progressive thickening of the parietal pleura. When it becomes extensive, it causes significant respiratory disability, as the lungs become effectively encased and immobile. High resolution CT or MRI scanning allows separate assessment of the pleura and the underlying lung parenchyma. In the absence of significant underlying asbestosis, there may be considerable improvement following surgical stripping of the pleura. Unilateral or bilateral diffuse pleural thickening, where the thickness extends to 5mm or more at any point, and involves more than 50% of the area of the affected chest wall in the case of unilateral disease, or more than 25% of the combined area of both lungs in the case of bilateral disease, is a prescribed industrial disease (PD D9).

Primary lung cancer

Asbestos fibres are known to increase the risk of bronchial carcinoma, at a level of exposure similar to that which results in a risk of asbestosis. There is also evidence of *synergy* between the risks of asbestos exposure and cigarette smoking. In other words, the combined risks are not merely additive, but multiplicative. In practice, this means that the combination of asbestos exposure and smoking creates a very high risk of lung cancer.

There is no way of distinguishing, either clinically or pathologically, between lung cancer caused by smoking, and that caused by other agents such as asbestos.

The symptoms may include:

(a) Cough

(b) Blood-stained sputum (haemoptysis)

(c) Progressive breathlessness

(d) Wheezing

(e) Chest pain

(f) Persistent chest infections

The examination findings may include

(a) Finger clubbing

(b) Wheezing

(c) Signs of infection

The diagnosis is usually confirmed by:

(a) Chest x-ray

(b) Bronchoscopy[40]

(c) Cytology[41]

(d) CT or MRI scanning of the chest.

If the tumour is diagnosed at an early stage, surgical excision may be possible, and is sometimes curative. In general, however, the disease has a poor prognosis. Other treatments such as radiotherapy or chemotherapy may improve survival, but are not curative.

Primary cancer of the lung is a prescribed industrial disease, where there is also evidence of occupational exposure to asbestos, and asbestosis, and/or diffuse pleural thickening (PD D8).

Mesothelioma

Mesothelioma is a highly malignant cancer, which usually arises in the parietal pleura.[42] The average latent period, following exposure to amphibole fibres, is about 40 years, but the total period or amount of exposure need not be great.

Rarely, mesothelioma can arise without any known occupational exposure to asbestos. It can therefore sometimes be very difficult to determine whether "guilty" or "innocent" fibres are to blame.

The presenting symptoms include:

(a) Breathlessness

(b) Aching chest discomfort.

40. Visual examination of the affected bronchus using a fibre-optic tube (a "bronchoscope"). Tissue samples may also be taken to confirm the presence of malignant cells.

41. Microscopic examination of sputum and samples of fluid washed out during bronchoscopy for malignant cells.

42. Mesothelioma sometimes occurs in the periotoneum or pericardium, following very heavy exposure to asbestos.

Examination usually reveals the presence of a pleural effusion. The investigations which are commonly undertaken include:

(a) Chest x-ray

(b) CT or MRI scanning[43]

(c) Pleural biopsy

(d) Cytology of pleural fluid[44]

(e) Thoracoscopy.[45]

There is no effective curative treatment, and treatment is therefore directed to palliative measures. These include:

(a) The instillation of drugs into the pleural space, in order to prevent effusions recurring

(b) Effective pain relief.

The prognosis is generally very poor. The average period of survival from diagnosis is about one year. Death is commonly due to blood clots in the lung (pulmonary embolus) or infections.

In some cases the diagnosis is uncertain prior to death, particularly where definitive tissue samples are not obtained.[46] The histo-pathological evidence used to establish the cause after death includes:

(a) The existence of changes of asbestosis in the lung parenchyma

(b) The presence of typical coated asbestos fibres ("asbestos bodies")

---

43. The typical findings are an irregular pleural tumour, with associated pleural effusion.
44. The diagnostic yield from pleural biopsy and cytology is fairly low. Thus it may be necessary to obtain a sample of tissue by means of direct visual inspection of the inner chest wall (thoracoscopy).
45. See note 44, above.
46. It is generally regarded as good practice to report all mesothelioma deaths to the coroner, as being due to suspected industrial exposure, even where occupational exposure is not suspected.

(c) Light microscope lung digestion. This may reveal high counts of straight uncoated asbestos fibres[47]

(d) Electron scanning microscopy may be used where uncertainty remains.[48]

Mesothelioma associated with occupational exposure to asbestos is a prescribed industrial disease, PD D3.

Prevention

When amphiboles are present, such as when old insulation is being removed, total protection suits are required, with an external air supply.

The standard method of monitoring personal exposure in the past relied upon examination of personal membrane respirator filters using an optical microscope. The exposure limit in the UK for chrysotile alone is 0.3 fibres/ml averaged over four hours, and 0.9 fibres/ml for a ten-minute period.

For any other forms of asbestos the limit is 0.2 fibres/ml averaged over four hours and 0.6 fibres/ml for a ten-minute period. However, this technique is probably inadequate for preventing mesothelioma, as most of the most toxic fibres will be invisible.

In the past heavy reliance was placed upon monitoring by means of chest x-rays. However, removal of the individual from further exposure once changes of asbestosis had appeared was unfortunately inadequate to reduce the long-term risk of mesothelioma and lung cancer.

Further Reading

Cockcroft A, Nemery B. "Inorganic Dusts" in: *Hunter's Diseases of Occupations* (9th edition). London, Edward Arnold, 2000.

47. Both techniques have disadvantages. Asbestos bodies tend to form around longer amphibole fibres, and may underestimate the exposure to shorter, more toxic fibres. Light microscopy of lung digests may underestimate the count of the thinnest fibres, and is unable to distinguish between chrysotile and amphibole fibres.
48. This technique is able to distinguish between chrysotile and amphibole fibres. Significant exposure is deemed to have occurred when counts of over one million amphibole fibres per dry gram of lung tissue are obtained, using digestion. Much higher counts are often found in mesothelioma cases.

Madan, I. "Occupational Asthma and Other Respiratory Diseases" in: *ABC of Work Related Diseases*. London, BMJ Publishing Group, 1997.

CHAPTER 7
WORK RELATED UPPER LIMB DISORDERS

SECTION 1: THE LEGAL PERSPECTIVE

Overview

The precise definition (and even in some cases the existence) of work related upper limb disorders (also referred to as "WRULDs") remains the subject of legal controversy.[1] The controversy lies in whether the condition actually results from a physical injury or is psychogenic in origin (or both). Most claims stem from office work, such as typing or VDU operation, but also from any repetitive industrial work for protracted periods. A typical example of this type of injury is shown by the facts of *Pickford* v *ICI Plc*:[2]

> "Miss Pickford was a diligent and conscientious employee. She was also an excellent and fast typist. In 1986 the typing content of her work was about 50 per cent. But it was her case at trial that by November or December 1988 that she began to experience strange feelings in her hands by the weekend. ..."

It is essential for the practitioner to have a clear comprehension of both the complex medical and the legal aspects of this type of claim. A summary of the proper approach in law to these cases is:

(1) Did the claimant employee suffer injury?

(2) Was the injury caused by any breach of duty by the defendant employer?

1. See for example the judgment of HHJ Prosser in *Mughal* v *Reuters* [1995] 6 Med LR 43.
2. [1998] 1 WLR 1189.

(3) What was the forseeability of the injury suffered as arising from the breach of duty?

(4) If so, can the cause of his/her injury be identified on the balance of probabilities?

The relevant case law

The facts of the main cases in this area of the law are summarised below. They form the basis of an assessment of the specific legal issues in each case, which are considered in greater detail in this chapter.

Mughal v Reuters Limited [1995] 6 Med LR 43

The claimant alleged he suffered injury in the course of his work as a journalist. He suffered pain in his fingers, arms, hands and shoulders as a result of the equipment he used. A central issue was whether repetitive strain injury ("RSI", but now known as "WRULD") actually existed. The claim was dismissed as the system of work was reasonably safe. Further, the pathology was more psychogenic than physical in origin, and moreover the judge concluded that "RSI" was not a recognised medical condition, namely one that had a distinct pathology.[3]

> "I have no doubt that ergonomists are well meaning in seeking to improve the comfort of people using visual display units. It is when RSI is alleged that it might be regarded as too easy a step for an ergonomist to supply the causation. That would be one uncertainty leading to another..."[4]

Pickford v ICI [1998] 1 WLR 1189

The claimant was a full time secretary and her duties included typing. In 1989 she complained of increasing pain in her hands and she sought medical advice. Her condition did not improve and she was dismissed on medical grounds in 1990. She contended that the cramp in her hands was caused by the breach of duty of her employer in failing to provide a safe system of work. Specifically she complained of having to type at high speeds for prolonged periods without proper rest breaks.

3. The defendant's medical expert in this case described the claimant/plaintiff's contentions as "ergonomic waffle!"

4. Note the contrasting findings in *Mountenay (Hazzard)* v *Bernard Matthews Plc* [1994] 5 Med LR 293.

The defendant employer contended that the claimant's injury was psychogenic in origin and so was not an "organic disorder" based on a physical injury.

At first instance, the claim was dismissed on all round unsatisfactory medical evidence. No organic pathology was proved, namely that the injury was caused by the typing as opposed to being merely associated with it.[5] Moreover, as to breach of duty, the employer was not under a duty to instruct the claimant to take proper rest breaks. The nature of her additional duties meant she had natural breaks from typing in any event.

The Court of Appeal reversed the judge's decision. On the medical evidence there was no evidence to show that her condition was not caused by the typing, which was persuasive of an organic cause. In addition an employer could not assume that an employee knows that breaks from typing are required and the omission to advise in this regard was negligent.

The House of Lords reversed the decision of the Court of Appeal. The decision was largely focused on the findings of fact.[6] As to negligence, it held the claimant failed, as the period of typing was not excessive and she had natural breaks afforded by her other non-typing activities in any event.

Alexander v Midland Bank [1999] IRLR 723

A number of part time bank encoders claimed that the system of work was negligent and caused injury. The job consisted of rapidly processing and recording details on cheques. The encoder would sit at a desk with the encoding machine. She would then pick up a cheque in one hand whilst at the same time keying in information into the encoder with the other hand. Minimum performance speeds were required and it was agreed that there were about 12,700 encodings per hour.

Additional relevant factual matters on liability were as follows:

(1) The system of work changed in that work was brought to their workstations and so the frequency and periods of breaks from encoding were reduced.

5. The medical causation issue is dealt with below.
6. Namely, that the claimant was found to be exaggerating her condition but this was not given sufficient weight by the Court of Appeal and so it was proceeding on incorrect factual assumptions.

(2) Output and performance were monitored and competition between the encoding centres was actively encouraged.

(3) By 1991 encoders suffered pains in the neck and the arms. The condition improved when they stopped the encoding, but ultimately they could not continue working.

The defendant bank accepted that the claimants were genuinely suffering symptoms of pain and disability, but contended that the symptoms were psychogenic and not physical. Further, that there were sufficient breaks from the encoding so as to prevent physical injury. At first instance judgment was given for the claimants. The evidence that the symptoms were psychogenic in origin was weaker than the contention for a physical cause. Poor posture and the almost constant repetitive tasks was compelling evidence on breach of duty and causation. The defendant appealed contending that the decision was ill-founded, namely by presenting the case as a choice between a psychogenic or physical cause of the injury. The contention was rejected and the employer's appeal was dismissed.

Gallagher v Bond Pearce (A firm) [2001] Lawtel 9/2/01 (Unreported)

The claimant worked as a legal secretary for the defendant solicitors' firm from 1986. She worked for a hard working partner and had a heavy workload. She used an electronic typewriter in 1990 and thereafter a word processor. An ergonomic assessment in 1992 indicated that she required more suitable work equipment. No action was taken and by 1994 she suffered pain in the shoulder, wrists and hands whenever she used the keyboard. In 1995 her work rota was revised and she had specific breaks from typing, but she still did five hours of typing per day. She was moved to a different department in 1995 but her employment was terminated at the end of 1995.[7]

The claimant alleged her WRULD was caused by the typing in addition to the inadequate nature of her work station. The claim succeeded based on common law and a breach of the Health and Safety (Display Screen Equipment) Regulations 1992. The defendant failed at the risk assessment stage to take the appropriate action with any urgency and her work system caused her injury. More precisely, the cause of the injury was

7. See also *Amosu* v *The Financial Times* [1998] 31.7.98 unreported.

multifactorial, but it was triggered by the breach of duty in relation to her work station and the heavy workload with which she had to deal.[8]

The Legal Principles: An Outline of Breach of Duty

Outline

A claimant has to prove a duty of care was owed in the course of the employment in question, namely to ensure the claimant employee was safe in his use of the system of work. This is usually not problematic as the duty arises from the employment relationship.[9] In addition a claimant has to prove a breach of duty by the defendant employer including that it was reasonably foreseeable that the injury would be caused as a result of the negligent system of work (not merely associated with it). Finally, the claimant must prove that the injury was caused, or materially contributed to, by the negligent system of work.[10]

Breach of statutory duty: the main regulations

Detailed consideration is needed of the particular statutory regulations in this area of the law. It not possible to cover all the regulations in detail but a summary is set out below:

(a) *Workplace (Display Screen Equipment) Regulations 1992*

- (i) *Regulation 2:* a suitable and sufficient analysis of the workstation has to be carried out for the purpose of health and safety risks.
- (ii) *Regulation 3:* the workstation has to comply with the standards set down in the schedule to the regulations.
- (iii) *Regulation 4:* work on display screen equipment has to be planned with periodic breaks, or changes in work activity.
- (iv) *Regulation 6:* training must be given prior to an audit by the employer.

8. See also: *Ball* v *Post Office* [1995] PIQR 5: where the claimant worked for the post office and contended that he suffered a "trigger-finger syndrome" from sorting letters. He did not claim under the Manual Handling Operations Regulations 1992 and the claim was made solely under the common law. The case is useful is showing the court's approach to the duty of care on the employer.
9. For further consideration of the duty of care see chapter 1.
10. Note that causation is considered in detail in chapter 2.

(v) *Regulation 7:* operators and users of relevant equipment must be given adequate information about all aspects of health and safety relating to work stations and all matters relating to the compliance with the regulations.

(b) The Management of Health and Safety at Work Regulations 1992 (as amended by 1999 regulations): regulation 3 requires a risk assessment.[11]

(c) Manual Handling Operations Regulations 1992: regulation 4, if lifting and carrying are involved in the injury.

(d) Workplace (Health, Safety and Welfare) Regulations 1992.

(e) Provision and Use of Work Equipment Regulations 1998.

(f) The Workplace (Health, Safety and Welfare) Regulations 1992: regulation 11 imposes a duty to provide a suitable seat at a workstation.

Health and Safety Publications

In support of any claim reference should also be made to the main HSE publications on the subject. These relate to the provision of a suitable work station, suitable training, warnings and a system of rotation of tasks. They are not legally binding but are persuasive. The more recent publications are:

(1) *Work related upper limb disorders: a guide to prevention* (1990) HSE

(2) *Upper Limb Disorders in the Workplace*: (2002) HSE.[12]

These publications and statutory codes provide guidance in assessing the knowledge of the risk of injury by the defendant, and whether a risk should have been appreciated and acted upon at common law. They are also useful in assessing what the proper practice was at the relevant time, but a breach of them is not necessarily negligent.[13] In fact care should be taken in

11. This does not impose any civil liability but it is relevant as to the evidence of a breach at common law.
12. Subsection 2(3) of the Health and Safety at Work Act 1974 requires most employers to have a written policy on health and safety issues. Section 16 HSWA 1974 provides for the provision of an approved code of practice.
13. See *Armstrong* v *British Coal Corporation* [1996] *Times* 6 December, as an example of where British Standards were relied upon.

over-relying on these publications as the primary basis of liability. In *Coley* v *Securicor Group Plc*[14] it was held that:

> "The real question is whether without the benefit of that (ergonomic) expertise the defendants should have perceived a sufficient potential risk to have at any rate sought such advice. All the plaintiff can rely upon in this regard are the 1997 and 1990 Health and Safety publications...Those to my mind fall far short of compelling the conclusion that reasonable employers would necessarily have been alerted to the risk."

The Breach of Duty: Assessment of the Case Law

The case law provides useful illustrations of the approach of the courts to various recurring themes in this area of the law.

Are there sufficient breaks and changes in activity?

In *Pickford* v *ICI Plc,* the House of Lords rejected the Court of Appeal's analysis on two bases:

(1) It wrongly assumed that the respondent employee's evidence, namely that she was typing for excessive periods without breaks, was in fact accurate.

(2) There was a failure to appreciate that alternative non-typing duties led to a "natural rotation" with the typing work, which was the finding at trial.

Nevertheless some further guidance on breach of duty was given. Comparison between the claimant's amount of typing and the accounts' department at her company (where there was extensive typing which justified regular breaks) found that the claimant was doing considerably less typing. Whilst forward planning was required for the accounts' department, in contrast the claimant could regulate her own work regime. Furthermore, even if the claimant had spent 75% of her time typing she still had 25% of her time to do other jobs, in addition to her lunch break, which was a critical distinction. The House of Lords held:

> "Unlike the typists she had both the experience and responsibility to organise and plan her own work according

14. [1998] Lawtel 3.2.98.

> to its requirements from day to day. She did not need to be told what to do."[15]

Consequently, there was no breach of duty by the employer in that:

(a) the claimant's actual working practices provided a natural break even if that was not specifically organised by the employer:

> "the nature of those other duties was such that she had to attend to them as and when they arose throughout her working day. They provided frequent natural breaks from typing as she answered the telephone, left her desk to speak to the managers elsewhere when they were not contactable by telephone, arranged meetings for them, made diary entries and so on."

(b) Her level of experience and responsibility was such that she could reasonably be expected to organise her own working time.

(c) The employer had taken steps to deal with the health and safety issues relating to work related upper limb disorders.

In *Alexander* v *Midland Bank* the defendant argued that 20 minute breaks in performing encoding each hour were sufficient to provide a reasonable break. The expert evidence agreed that point in principle, but the Court of Appeal found that in fact there were no such breaks. Reliance was placed by the defendant upon data obtained from the encoders that recorded "dead time", but no expert evidence was called in relation to this issue. The judge at first instance held that no reliable pattern could be inferred[16] and so the claimants' contention of breaks of only eight minutes per hour was accepted. Further, the competitive atmosphere encouraged by the employer meant that the employees were always in a tense situation and pushed beyond their capabilities. Hence the defendant was well aware of the risk but had not taken reasonable steps to reduce or eradicate it and so the appeal failed.

15. [*Supra*] at 1206A.
16. The inference from the judgment is that the failure to have a witness explain the data relied on was fatal to the success of the point.

An assessment of the employee's ergonomic requirements

In *Gallagher* v *Bond Pearce*: the claimant employee contended that her heavy typing load was the cause of her injury. The defendant employer was found in breach of the Workplace (Display Screen Equipment) Regulations 1992, but a central finding was that no proper audit of the claimant's ergonomic requirements had been made and no action was taken within a reasonable time. In contrast, in *Pickford* v *ICI* it was held that the employer had made a serious and sufficient attempt to address the health and safety requirements of the employees and these had been addressed properly.

Employers have to act on any ergonomic assessment and there is a continuing duty to provide proper and safe equipment. A failure to do so will probably be held as a breach of duty, although any such breach must be causative of the alleged injury. All cases are fact sensitive but material factors include:

(1) The frequency and duration of the work activity.

(2) The force applied during the work activity

(3) The posture of the employee whilst at work.[17]

In *Alexander* v *Midland Bank*, there was criticism of the body posture of the claimants at their desks and in relation to their hands on the encoders. They were described as "sitting like bags of coals" but the defendant employer attached little importance to posture until 1992 when guidance was given.

Should a warning be given?

Employers are under a duty to provide instruction to the employee on all matters relating to health and safety, pursuant to the Display Screen Equipment Regulations 1992, as in *Gallagher* v *Bond Pearce*.[18] Any such instruction must be sufficient to discharge the duty of care such that a reasonable employee would have acted upon it. The duty is a continuing

17. The HSE publication *Work Related Upper Limb Disorders* 1990: stated the relevant factors were force, frequency and posture and the particular job has to be considered in that context.
18. See also *Mountenay* v *Bernard Matthews plc* [1994] Med Law Rep 293.

one throughout the relevant period of employment.[19] Whether an employer should remove or dismiss an employee who knows the risk of injury but wishes to run it anyway, is dependant on the magnitude of the risk.[20]

Nonetheless, a distinction has to be drawn between the requirement of proper instruction and training of employees and the more generalised allegation of a failure to warn employees of the risks of injury related to the job. In *Pickford* v *ICI* the House of Lords considered the issue of a warning of the risk injury. Whilst it accepted the need to warn in principle it was a limited and qualified acceptance:

> "The giving of warnings of the risk of disease or injury is a precaution which is familiar in the field of litigation for personal injury. But in the case of conditions such as PDA4 which are not easily identifiable and not well understood, great caution must be exercised as to the content of any such warning and as to whether to give a warning at all is appropriate. To impose a duty which may cause more harm than good would be undesirable."[21]

The duty to warn has to be considered on the facts of the particular case. First, if the employee is sufficiently experienced and able to organise the system of work such that he has suitable breaks then the relevance of a warning on the point is questionable.[22] Courts will take a common sense approach to the relevance and efficacy of any warning, specifically whether the task performed by the employee required one. A risk:benefit analysis has to be performed by the employer, but if the employee would not have acted upon any warning in practice then causation cannot be made out. In *Pickford* v *ICI* the expert evidence supported the contention that in 1989:

> "A formal system of instruction, warning and advice was adopted and implemented for the typing staff in the accountant's department as the working day was confined to accounts and difficulties in changing postures could arise in their case. But such a system was not considered

19. Note that medical opinion is not unanimously of the view that once symptoms present then the employee should be removed from the task, which is relevant to the use of warnings in practice.
20. *Coxhall* v *Goodyear Great Britain Limited* [2002] EWCA Civ 1010.
21. [*Supra*] at 1206.
22. A point emphasised in *Pickford* v *ICI*.

necessary for secretaries as they carried out many non-typing duties in the course of their working day."

The particular system of work

There is a broad requirement to provide a safe system of work, but analysis is required of the particular activities undertaken by the claimant employee. For example, in *Pickford* v *ICI* and *Gallagher* v *Bond Pearce,* close assessment was made of the amount of typing undertaken by the employee, the extent of the breaks, the equipment provided. In *Alexander* v *Midland Bank* a central issue was the amount of keystrokes on the encoders by the bank staff. It was held that if an employee suffers pain and loss of function and reports this to the employer, then there is a duty to change the nature of employment. Dismissal depends on the magnitude of risk but also the practicability of the employer being able to provide alternative employment. It is emphasised that employers only need to take reasonable steps to ensure the system of work is safe. If the employer has addressed the health and safety issues and the system has been varied and/or rotated and/or the type of jobs enlarged then it is unlikely that a breach of duty will be made out.

Foreseeability

A claimant has to prove that the type of work that he was undertaking entailed a foreseeable risk of injury, whether physical or psychiatric. The extent of injury is not a relevant factor. First, it has to be shown that the risks of injury were known, or ought to have been, known by the defendant at the material time, or this will provide a full defence on the issue of foreseeability. For example in *McSherry* v *British Telecom*[23] it was held the defendant should have been aware of risks of upper limb disorders from keyboard use by 1985.[24]

As to foreseeability of injury itself, in *Pickford* v *ICI* the Court of Appeal held it was reasonably foreseeable that typists might suffer injury if they typed for long hours without a break. Whilst the House of Lords reversed the decision of the Court of Appeal as it proceeded on unjustified factual assumptions, there was no specific rejection of the principle. Hence, the point has

23. [1992] 3 Med LR 129.

24. In *Graham* v *Coop Wholesale Society* [1957] 1 All ER 654, it was held there is a continuing duty on the employer to keep up to date with health and safety issues.

to be assessed on the facts of the individual case and it is a matter for the trial judge. The House of Lords held:

> "The findings by the judge that the condition was not reasonably foreseeable in her case and that the appellants were not negligent in the respects alleged by her were in my opinion sound based on the evidence."

Gallagher v *Bond Pearce*, held that it was reasonably foreseeable that the claimant would suffer an injury if insufficient attention was paid to such factors as the work station, the volume of work and the provision of rest breaks. As the defendant knew of the health risks from keyboard work the claimant did not have to prove that actual injury was foreseeable, but merely that the damage was of a similar kind to that which was foreseeable, such as cramp or tenosynovitis. Moreover, foreseeability was made out on the basis that the defendant had settled previous similar cases before this one. In addition, the knowledge of the risk known to employer was founded upon the various publications from the Health and Safety Executive in the 1980s on the problems relating to keyboard operations. The extensive publications on this issue makes it difficult for an employer to argue that it was not aware of the risk, although such knowledge would not, of itself, make the risk foreseeable.

Causation

Has an injury been proved?

The claimant has to prove a recognised injury, that was caused by the system of work as opposed to being merely work-related. Whether there has been an "injury" can be controversial. There is no problem with a recognised clinical condition (e.g. carpal tunnel syndrome) but problems arise with more non-specific pain syndromes and it will be a matter for the judge to assess whether an "injury" has been suffered that sounds in damages.

As to causation of the injury, the issue is not whether the work has caused the injury to the employee, but whether the alleged breach of the employer's duty of care caused the injury. In general, if the claimant proves there has been a breach of duty, and that breach makes a material contribution to the injury then he succeeds.[25] Causes can be multifactorial and the

25. *Bonnington Castings Limited* v *Wardlaw* [*supra*].

claimant only has to succeed on one of them. Nevertheless the claimant has to prove his case and if the court is uncertain about the cause then the claim must fail.[26]

What Injury is in Issue?

A central issue on causation is whether the alleged injury suffered resulted from physical injury (i.e. an organic pathology) or was psychological in origin. The fact that a claimant has suffered an injury which is psychogenic in origin, as opposed to physical, does not mean that causation is not made out.[27] If psychiatric injury is the reasonably foreseeable consequence of the breach of duty then damages are recoverable. But if the injury is psychogenic, the practical difficulty is not that the damages are irrecoverable, but that causation cannot be made out between the breach of duty and the injury.[28] Thus an important causation issue between the employer and employee will be whether there is an organic pathology or whether the symptoms are psychogenic in origin.

A point of entry on causation is to assess the type of injury suffered. The case law is analysed below:

(a) *Pickford* v *ICI*: the injury was cramp in the hands allegedly caused by typing. There were two competing explanations for the condition on the medical evidence, either psychogenic or physical/organic (PDA4).

(b) *Alexander* v *Midland Bank*: the injury was musculoskeletal to necks, arms and hands. There were two possible explanations, a physical one and a psychogenic one. Competing medical evidence for each school of thought had to be assessed.

(c) *Mughal* v *Reuters*: the injury was either organic or psychological and it was held the symptoms were an unassociated psychogenic condition.

(d) *Mountenay* v *Bernard Matthews*: the claimants suffered from tenosynovitis-like symptoms. The injuries were beyond normal aches and pains and fell short of actual injury, but the claimants nevertheless succeeded. No wider diffuse condition was proved but the abnormal pain was caused by the work conditions.

26. *Gallagher* v *Bond Pearce* [*supra*].
27. *Page* v *Smith* (1996) [*supra*].
28. See Smith LJ: *Alexander* v *Midland Bank* [*supra*].

(e) Generally, where the symptoms are more diffuse then the claimant will probably find it more difficult to prove causation.

(f) Finally, an allegation of malingering provides a possible causal explanation for the presentation of the alleged symptoms.

Assessment of the Courts' approach to causation

In *Pickford* v *ICI* the competing medical explanations were psychogenic, as argued by the defendant and organic, as contended by the claimant's experts.[29] The House of Lords did not consider either medical view wholly satisfactory. For example, the claimant had poor function in the hand, but her grip was relatively normal which tended to undermine an organic cause. But the defendant's assertion of a conversion hysteria was also undermined by other expert evidence. The House of Lords endorsed the trial judge's rejection of an organic cause; a conclusion drawn from his assessment of the conflicting medical and ergonomic evidence. Effectively, the claimant's evidence was rejected as being exaggerated and so she had not proved her case. There was a specific rejection of the approach of the Court of Appeal, which stated that once the defendant's contention of a psychogenic cause was rejected it automatically meant the organic pathology was proved. The rejection of the defendant's case did not necessarily mean the claimant had discharged the burden of proof. It was only a relevant factor to take into account.

In contrast in *Alexander* v *Midland Bank* the issue was whether the judge at first instance had:

> "taken the wrong approach in concluding that since the evidence did not establish that the condition was psychogenic in origin the probability must be that it was physical."[30]

At trial, the judge concluded there was no substantial evidence in favour of psychogenic causation. In consequence, the evidence in favour of the physical cause must be "strong",

29. An interesting point raised in *Alexander* v *Midland Bank* [*supra*] was that an ergonomist who favoured the psychogenic school of causation would have some difficulty in commenting on causation in the context of the layout of the workstation as on the psychogenic assessment this could not be causative [see [1999] IRLR 723: at para 20].

30. [1999] IRLR at 723.

particularly as there was no allegation of malingering. The Court of Appeal accepted the strict legal test on the burden of proof but stated:

> "...in practice where the parties argue for one of only two explanations who can prove what to be the more probable must necessarily be an important factor in evolving the final solution."[31]

The Court of Appeal made additional observations on the competing causation theories of the medical experts:

(a) The psychogenic position argued that patients can appear outwardly normal but they have picked up psychosomatic disorders, possibly from work colleagues.

(b) The defendant's orthopaedic surgeon asserted that the rheumatological analysis was riddled with holes such that the only feasible explanation was that the symptoms were psychosomatic. However, the orthopaedic expert did not have the expertise to comment on psychological issues and this undermined his evidence.

(c) The defendant contended for a psychogenic causation, but this argument was weak in the context of a group of claimants. The characteristics sought to be applied by the medical expert in support of the case did not fit with the characteristics of the conventional hard working claimants who gave evidence.

The decision in *Alexander* v *Midland Bank*, substantially rejected the contention of a psychogenic cause, but superficially this does not sit happily with the House of Lords' decision in *Pickford* v *ICI*. Ultimately it may be matter of emphasis, as the House of Lords did not reject the approach but indicated that the rejection of one school of medical thought was a "factor" in assessing causation. It is probably an issue of weight as opposed to principle, but with the burden of proof remaining squarely on the claimant.

In *Gallagher* v *Bond Pearce* the defendant contended there was no organic pathology that explained the diffuse symptoms. Neither expert considered that the claimant was a malingerer,

31. At para 32. IRLR 729.

but a distinguishing feature was that there no suggestion that the symptoms in this case were psychogenic in origin. The issue was therefore whether there was an organic pathology. The defendant's expert made the following points:

(1) there were thousands of people doing typing work who never complain of these symptoms.

(2) There is nothing inherently dangerous in typing.

(3) Non specific pain was prevalent in the population at large, including the working population.

(4) But he also accepted that he could not say what the cause of her symptoms was and so causation remained undecided to that extent.

The judge was satisfied that the claimant had suffered an injury in the form of diffuse aches and pains. This had been caused by the excessive workload, poor posture and insufficient breaks from working.[32] If the defendant had taken prompt action then injury may have been averted, but no action was taken. The judge specifically held that medical controversy did not preclude a final judgment on the point. The trigger was the overtime, but the causes were multifactorial. The claimant had suffered a work related upper limb disorder and her expert evidence was preferred to that of the defendant, who could provide no explanation at all for her continuing symptoms.

Finally, as stated above, there is no need to prove that the breach was the only cause as long as the breach relied upon made a material contribution to the injury. But there is still the overriding issue of apportionment as set out in *Holtby* v *Brigham and Cowan Limited*,[33] namely that where the injury is "divisible" then the defendant is liable only for that amount of the injury caused by the negligence. The correct approach to the assessment of damages is set out in *Sutherland* v *Hatton*.[34]

32. That is, the condition went further than the additional "aches and pains" caused in *Mountenay* v *Bernard Matthews*.
33. [2000] 3 All ER 321.
34. [2002] IRLR 263: the issue of causation and apportionment is covered in Chapter 2 on causation.

Summary and Practice Points

Practitioners need to have a clear appreciation of the issues in this type of case. A summary of important points and issues for both defendant and claimant practitioners is set out below.

Issue	Claimant's Perspective	Defendant's Perspective
What is the nature of the injury What type of WRULD is it?	• Assess the nature of the claimant's injuries. Is it an "injury", i.e. a recognised medical condition based on an organic cause? • Ensure the diagnosis is made on basis of all clinical notes. Consider limitation.	• Make a similar assessment and obtain the medical notes. • Ensure the diagnosis is based on all those notes. Consider limitation.
What is the evidence needed on liability? This is based substantially on statutory regulation: Workplace (Display Screen Equipment) Regulations 1992: Management of Health and Safety at Work Regulations 1999.	***On liability*** • Ergonomic expert evidence is required. • Assessment of the system of work, including comparison with other work colleagues. • Seek full disclosure on system of work.	• Same as claimant. Make early investigations on the system of work and the colleagues.
Evidence: What are the particular evidential requirements? (a) System of work, amount of typing etc. (b) Breaks from work. (c) Seating/workstation. (d) Onset of symptoms.	***On breach of duty*** • Key issues are work rates, frequency of breaks, natural breaks, force, frequency, duration: refer to HSE publications. • Was the claimant aware of the health and safety issues?	• Same as claimant. Obtain by Part 18 request the claimant's precise case on breach of duty at early stage.

Issue	Claimant's Perspective	Defendant's Perspective
Foreseeability: The issue on foreseeability is central, including the date of knowledge of the risk of injury.	• Did the employer know of the risk of injury? What are the risk assessments, previous cases, knowledge of HSE publications? Assess working practices and health and safety issues of employer. Are warnings relevant? What was the claimant's own control?	• Assess the employer's health and safety track record. • Identify the knowledge of the employer of this risk of injury. Assess breach of statutory duty. Was it causative? Consider warnings and notification by claimant of symptoms.
Causation In fact and other evidence: the system of work must have caused the symptoms.	• Was the work system causative of the injury or just associated with it? Identify the connection. • What was the first date of presentation of symptoms?	• Causation: has the breach of duty caused the injury? How? • Identify the dates when symptoms first presented and the nature of the allegations. • Are the two inconsistent?
Experts: What are the related medical evidential issues? Expert evidence must cover breach of duty and causation.	• Do the medical records show when the symptoms first presented?	• Assess the medical records specifically on the symptoms and the trigger date.

Issue	Claimant's Perspective	Defendant's Perspective
Medical/Ergonomic expert evidence: It is vital to show that some injury was caused by the system of work and also medical causation.	• What sort of WRULD case is this? organic/psychogenic, or merely organic? Is there malingering? Is it multifactorial? Select the appropriate expert. • Who is the correct expert? [Orthopaedic rheumatological] • Is there a psychiatric element? Get the expert evidence. • Experts: to state precisely the injury and causation. • Has an "injury" been established? i.e. the type of WRULD case. • Experts must provide medical authorities.	• What is cause of the claimant's injury: organic psychogenic or malingering, or merely no pathology at all? • Ensure that the correct expert speaks to these particular conditions. • If the defence is a psychogenic pathology, what is the relevance of the breach of duty? • If relevant what does ergonomist say on causation? • Expert must examine the medical records on causation. Is the claimant a malingerer or exaggerator? • Can the medical expert attribute a cause at all to the symptoms?[35]

35. If the expert's evidence is that there is no cause of the symptoms and it cannot be categorised and in contrast the claimant's expert is clear that the symptoms are caused by the system of work then all things being equal the court will probably find for the claimant: (see *Gallagher*).

Quantum of General Damages

These types of case are fact sensitive but a broad indication can be gained from the case law set out below. As to the approach to the calculation of damages reference should be made to *Sutherland* v *Hatton*[36] which is fully discussed in the chapter on causation.

Name of Case	**Facts and Award**
Gallagher v *Bond Pearce* [2001] Current Law 1668	Female secretary, 45 employed from 1986 to 1995. In 1995 developed stabbing pains in the shoulder. These gradually deteriorated and spread to her hands and wrists when she was using a keyboard. She was dismissed in November 1995. She could not return to typing work. *GD: £7,300*
Hunter v *Clyde Shaw Plc* [1995] S.L.T. 474	Industrial radiographer, 48 developed lateral epicondylitis (tennis elbow) from moving castings mainly of between one and five tonnes at work. Made redundant in December 1991. He underwent surgery on his right arm but was left with weakness and pain in addition to reduced grip. *GD: £8,000*
Molvaer v *Whirly Bird Services Ltd* [2000] Current Law 1605	Female, 41 was employed to service and repair diving suits for the offshore industry. She suffered an overuse syndrome in both upper limbs which caused bilateral pain, exacerbated by manual tasks. Gripping was difficult and she could not be employed. Household chores and her hobbies were affected. She was expected to recover after six months of active management. *GD: £7,000*

36. [*Supra*].

Name of Case	Facts and Award
Ping v *Eselte Letraset Limited* [1992] PIQR P74	A series of claimants was awarded varying sums for injuries caused whilst performing factory work. They suffered from varying complaints including carpal tunnel syndrome, tennis elbow, trigger thumb and golfer's elbow. The awards were in the region of £3,000–£8,000 and these were upheld on appeal.
Smith v *Baker and McKenzie* [1994] Current Law 1662	Female secretary suffered injury as a result of a negligent system of work between 1988 to October 1990. She suffered left sided tenosynovitis and stiffness of the fingers with numbness of the wrist and knuckles. *GD: £7250*
Hook v *Honeywell FM Ltd* [2001] Current Law 1667	Female, aged 27, employed as a secretary. Increase in typing duties up to eight hours a day. She suffered shooting pains in her dominant left arm and was diagnosed with chronic compartment syndrome in the left arm. She underwent surgery to release pressure in the forearm, which resulted in minor permanent scarring on the forearm. The claimant would be unable to type again in the future but she made an 85% recovery from her injury. *GD: £11,000*

SECTION 2: THE MEDICAL PERSPECTIVE

Overview

Work-related musculo-skeletal disorders probably represent the largest category of work-related illness in the UK. Many of the affected individuals have symptoms confined to the upper limbs and neck. For this reason work-related neck conditions are also dealt with in this section.

"WRULD" is not a medical diagnosis. It is an umbrella term used to describe symptoms of pain, stiffness, weakness, swelling and sometimes sensory disturbance in the soft tissues of the fingers, hands, wrists, arms and shoulders, due to a wide variety of work-related conditions. Over recent years many other terms have been used synonymously with WRULD, including repetitive strain injury ("RSI"), cumulative trauma disorder ("CTD") and overuse syndrome.

These terms have also been used synonymously with upper limb disease terms such as carpal tunnel syndrome, tennis and golfer's elbow, tenosynovitis and de Quervain's syndrome.

Basic Anatomy

The principal movements of the fingers are flexion (bending) and extension (straightening). Movements of the thumb are more complex, and also include abduction[37] and adduction.[38]

The flexor and extensor muscle compartments form most of the bulk of the front and back of the forearm respectively (see figure 3). The tendons, which attach the muscles to the bones, run in sheaths lined with synovial membrane.[39] The flexor tendon sheaths extend through the palm to the base of the fingers and thumb. The extensor tendon sheaths do not extend beyond the wrist.

37. Movement away from the palm, in the same plane as the palm.
38. Movement towards the palm, in the same plane as the palm.
39. The same type of membrane as that which lines and lubricates joints.

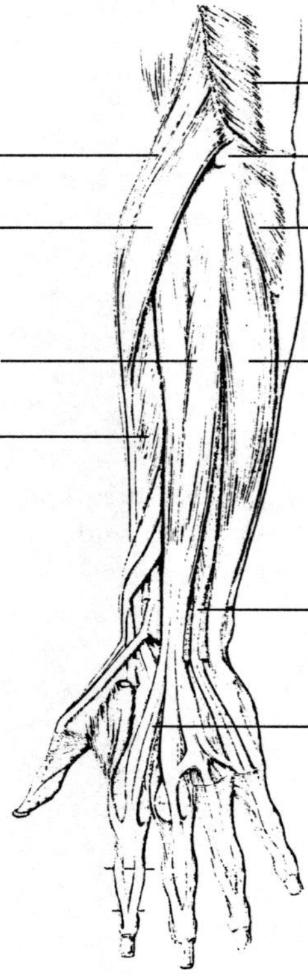

Figure 3
Posterior Surface of Fore-arm

The principal movements of the wrist are flexion and extension. At the elbow, in addition to flexion and extension, the rotational movements pronation and supination[40] are also important.

The shoulder is a shallow ball and socket joint, which permits a wide range of movements including flexion, extension, abduction, adduction, and rotation about the long axis of the arm.[41]

The "rotator cuff" is the name given to the sheath of tendons of the short shoulder muscles,[42] which cover and blend with the joint and its capsule, and provide most of its mechanical support.

40. Pronation is rotation of the forearm so that the palm faces downwards, supination is the opposite movement, so that the palm faces upwards.
41. However, movements at the shoulder joint itself cannot be divorced from those of the shoulder girdle as a whole, which also include the clavicle (collar bone), and the scapula (shoulder blade).
42. Supraspinatus, infraspinatus, subscapularis and teres minor.

Common Conditions

Tendinitis and tenosynovitis

Tendinitis is inflammation of tendons. Tenosynovitis is inflammation of tendon sheaths. The conditions often occur together. The typical symptoms are pain, discomfort and difficulty using the affected muscles and their tendons.

The extensor tendons of the fingers and their tendon sheaths are commonly involved following arduous or unaccustomed manual work, or frequent or repeated movements of the hand or wrist. The site of the pain and discomfort is on the back of the wrist, where the tendons run through the mouths of their sheaths. Examination may reveal tenderness and swelling, usually confined to the area of the tendon sheaths on the back of the wrist. Sometimes "crepitus" is also felt.[43]

Industrial workers or crop production workers exposed to tasks that require use of the hands above shoulder height, such as reaching, lifting, and repeated use of the arm in abduction and flexion, are vulnerable to "rotator cuff tendinitis". This usually involves the supraspinatus tendon.[44]

The typical symptoms are dull aching shoulder pain aggravated by reaching up, dressing and overhead work. Examination may reveal tenderness over the head of the humerus. Pain is sometimes reproduced by movement of the affected tendon against resistance, and there may be a "painful arc".[45]

Traumatic inflammation of the tendons of the hands or forearm, or the associated tendon sheaths, is a prescribed industrial disease (PD A8), when caused by arduous or unaccustomed manual work, or frequent or repeated movements of the hand or wrist.

43. A creaking sensation over the affected area, as the tendons move in their sheaths.
44. Where the head or ball of the humerus meets the joint between the clavicle and a bony protruberance of the scapula known as the acromion (the acromio-clavicular joint), the supraspinatus tendon passes through a narrow tunnel. Repeated elevation may compress the tendon at this site, causing tendinitis through a combination of mechanical friction and interruption of the blood supply.
45. This occurs during abduction, between 70 to 120 degrees, as the sensitive part of the tendon passes through the tunnel.

Trigger digit

This is a specific form of tendinitis involving the flexor tendons or their sheaths. A finger or thumb may catch, or even become stuck in a flexed position ("triggering"), as the swollen tendon impinges at the mouth of the tendon sheath. This condition is commonly seen with repeated gripping activities of the hands or fingers, such as overuse of pistol air tools.

Common sites for painful catching or triggering are at the base of the fingers where the palm is creased (near the metacarpo-phalangeal joints), or at the joints between the bones of the thumb (the inter-phalangeal joints). Examination may reveal clicking, a "pop", or crepitus, where the tendon enters its sheath.

This condition falls within category PD A8 of the prescribed industrial diseases, when caused by arduous or unaccustomed manual work, or frequent or repeated movements of the hand or wrist.

De Quervain's syndrome

This is a special form of tenosynovitis involving only two of the tendons of the thumb, the abductor pollicis longus and the extensor pollicis brevis. It is associated with overuse of the thumb, and repeated ulnar or radial deviation of the wrist, such as using a pinch grip or grasping, for example when wringing out cloths.

In this condition dull aching pain, and sometimes clicking, occur at the radial (outer) border of the underside of the wrist, where the tendons pass under a tight fibrous bridge. Examination may reveal tenderness, a characteristic sausage-shaped swelling, and crepitus at this site.[46]

This condition falls within category PD A8 of the prescribed industrial diseases, when caused by arduous or unaccustomed manual work, or frequent or repeated movements of the hand or wrist.

46. A clinical test known as *Finkelstein's test* is sometimes employed to assist with diagnosis. With the thumb tucked tightly into the fingers, making a fist, the wrist is gently pushed into flexion and towards the inner (ulnar) side, which stresses the tendons and reproduces the symptoms. However, the test is not wholly reliable, and may be positive in the absence of true De Quervain's syndrome, due to other conditions such as arthritis.

Carpal tunnel syndrome

On the underside of the wrist, a trough formed by the bones of the wrist below, and a fibrous roof above (the flexor retinaculum), provides a tunnel for the median nerve and the long flexor tendons and tendon sheaths of the fingers and thumb (see figure 4). There is no room in the tunnel for the tissues to expand, and swelling of the tendons, or their sheaths, or other soft tissues at this site may compress the median nerve. This gives rise to the symptoms of carpal tunnel syndrome ("CTS").

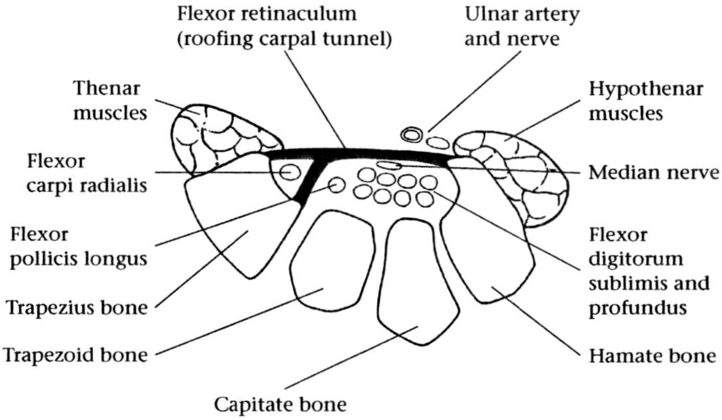

Figure 4
Transverse section through the carpus, showing the attachments of the flexor retinaculum

The symptoms are:

(a) Pain, tingling and numbness, or any combination of the three, involving the palmar surface of the thumb, index and middle fingers, and usually the lateral (radial) half of the ring finger.[47]

(b) Over time the pain may extend up as far as the elbow.

(c) The symptoms are often worse at night, and may disturb sleep.

47. The little finger is never involved. The palm itself is not usually involved, since the palmar branch of the median nerve arises above the wrist and is not trapped in the tunnel.

Examination may reveal:

(a) Impaired sensation in the affected area.

(b) Sometimes there is also weakness of the muscles innervated by the median nerve, including some of the thenar muscles, which form the fleshy "thenar eminence" of the palm, at the base of the thumb.

Clinical tests which are employed include:

(a) *Tinel's test*; reproduction of symptoms by tapping the median nerve where it lies in the carpal tunnel.

(b) *Phalen's test*; reproduction of symptoms by holding the wrist in 90 degrees of flexion for sixty seconds.

(c) *Nerve conduction studies*, which measure conduction of electrical impulses across the affected nerve, are diagnostic, but usually only when the condition has been present for some time.

CTS commonly arises due to conditions unrelated to work, for example due to fluid retention in pregnancy or after prescription of the oral contraceptive pill, and due to soft tissue thickening in conditions such as rheumatoid arthritis, gout, diabetes mellitus and hypothyroidism (lack of thyroid hormone).

It may occur in the industrial setting with repeated flexion or extension of the wrist, particularly where excessive force or vibration is also involved. However, CTS is only a prescribed industrial disease when associated with hand-held vibrating tools (PD A12), or when it can be shown that it is a direct consequence of tenosynovitis of the finger flexor tendons in the carpal tunnel.

Peritendinitis crepitans

Sometimes tendon inflammation is more proximal (nearer to the elbow), at the junction between the muscles and their tendons around the middle of the forearm. This mild variant of tendinitis is known as "peritendinitis crepitans". As with tendinitis and tenosynovitis, it more commonly involves the extensor muscles and their tendons.

The symptoms are similar to those of tendinitis/tenosynovitis, but are felt towards the middle of the forearm. Examination may reveal tenderness in the affected area, and sometimes a characteristic fine crepitus.

This condition falls within category PD A8 of the prescribed industrial diseases when caused by arduous or unaccustomed manual work, or frequent or repeated movements of the hand or wrist.

Epicondylitis

The areas of insertion of the tendons proximally may also become inflamed. "Tennis elbow" is inflammation in the region of the lateral (external) condyle of the elbow, where the common extensor muscle tendons attach to the bone (see figure 3).

The symptoms are aching over the lateral aspect of the elbow, particularly when gripping strongly, rotating the forearm or extending the wrist.[48] In the industrial setting, repeated pronation and supination, forceful wrist extension and flexion, or grip with a fully extended or flexed arm may be more important.

Examination may reveal tenderness over the lateral condyle. Cozen's test involves stressing the extensor tendon origin, by forced resistance of wrist extension and radial deviation, while the forearm is pronated. This may reproduce the symptoms.

"Golfer's elbow" is inflammation in the region of the medial or inner condyle of the elbow where the common flexor muscle tendons attach. The symptoms and examination findings are similar to those of tennis elbow, save that they are localized to the region of the medial (inner) condyle of the elbow, and may be reproduced by forced resistance of wrist flexion.[49]

Epicondylitis is not a prescribed industrial disease. It is also a common complaint in the non-working population.

48. Classically, the symptoms are provoked by strong flexion of the wrist while the extensor muscles are contracted, as in the backhand stroke of tennis.
49. Classically, the symptoms are precipitated by a golfer striking the ground instead of the ball.

Writer's cramp

Writer's cramp[50] is a condition which results in pain, muscle tremor and muscle spasm in the hand or forearm due to repetitive movements of the fingers, hand or arm. It has been associated with prolonged periods of handwriting, typing and other repetitive movements, such a playing musical instruments.

The question of whether it is a true medical entity is disputed. Some experts believe that it is a combination of the normal reversible physical fatigue of muscles and an underlying psychological condition. It is a prescribed industrial disease (PD A4), although it is diagnosed infrequently in the workplace.

Neck disorders

These are sometimes referred to as "cervical syndrome" or "tension neck syndrome". Cervical syndrome is said to be present when pain in the neck is associated with radiation of pain along a cervical nerve root.[51]

The typical symptoms of tension neck syndrome are fatigue, aching and stiffness in the neck. The pain may radiate into the occiput (back of the head), or into the shoulder and upper limbs. There may be associated muscle tension headaches, due to spasm of the trapezius and para-vertebral muscles. Sometimes there are palpable tender spots in the muscles.

It has been associated with prolonged static flexion of the neck, such as when a visual display unit is poorly positioned, or extreme repeated twisting movements of the neck. Sometimes x-ray examination reveals degenerative disease of the cervical spine.[52]

The condition is common in the non-working population, and is not a prescribed industrial disease.

---

50. Known historically as telegraphist's cramp, twister's cramp, occupational or professional cramp and craft palsy. It was first described by Ramazzini in 1713 in scribes and notaries.
51. It may therefore be indistinguishable from the symptoms of cervical degenerative disease, which is common in all adults from middle life, irrespective of their occupation. In the absence of evidence of true cervical degenerative disease involving nerve roots, it is probably a form of regional fibro-myalgia, localized to the neck and shoulder area (see below).
52. However, symptoms correlate very poorly with the presence or absence of such x-ray changes, and many patients with significant radiological degeneration are free of symptoms.

The "beat conditions"

Repeated use of picks, shovels and hand tools has sometimes been associated with a sterile subcutaneous cellulitis (inflammation of soft tissues without infection) of the hand ("beat hand").

Physical trauma, friction and pressure are thought to be causal factors. The symptoms are pain, tenderness, redness and swelling. A sterile inflamed bursa (bursitis) or cellulitis may give similar symptoms at the elbow ("beat elbow") and the knee ("beat knee"[53]).

All three conditions are prescribed industrial diseases when associated with severe or prolonged friction and pressure at the site of the affected part: beat hand is PDA5, beat knee is PDA6, and beat elbow is PDA7.

Regional fibro-myalgia

Regional fibro-myalgia, sometimes referred to as "diffuse RSI", "non-specific RSI", "non-specific arm pain" and "myo-fascial pain syndrome", is a condition in which pain occurs, in varying distribution and degree, usually involving the neck, shoulder, arm, elbow, wrist or hand. This is sometimes associated with localized tender spots, and abnormal nerve tension and grip tests.

The condition is controversial, and some experts dispute whether it is a real disease entity at all.[54] The points against its existence are:

(a) There is no demonstrable patho-physiological basis for the symptoms, and no generally accepted or identifiable pathological lesion.

(b) The reported pain may not follow a recognised anatomical or neurological distribution.

(c) Tests for localised tenderness, grip and nerve tension are unreliable, as they depend upon subjective factors such as the degree of pressure applied and the patient's reaction. False positive reactions may occur in healthy individuals.

---

53. Beat knee may be associated with prolonged kneeling, such as in mining, carpet laying and plumbing.
54. Anecdotal evidence suggests that the condition is more readily diagnosed by rheumatologists and occupational physicians, and less readily diagnosed by orthopaedic surgeons.

(d) Many of the available research studies are flawed by the absence of adequate control groups.

(e) The condition is therefore probably no more than the discomfort ordinarily associated with reversible muscle fatigue caused by repetitive activity or prolonged static posture, or a psychosomatic condition such as compensation neurosis or conversion hysteria, or a combination of both.[55]

The points for its existence include:

(a) The symptoms may emanate from local tissue damage, possibly involving muscle fibres, blood vessels or nerves.

(b) Some studies have reported abnormalities of certain muscle fibres and their micro-circulation, following prolonged low static loading of muscles such as the trapezius at the neck and shoulder.

(c) Abnormally sensitised pain receptors in the spinal cord or brain may play a role.

(d) The finding of localised tender spots is objective evidence of tissue damage or injury, just as valid as an abnormal reflex in a nerve injury.

(e) In the absence of such objective signs, pain and appropriate occupational exposure may suffice to make a diagnosis.

(f) The mere fact that there is currently no precise physiological or pathological explanation does not mean that the condition is all in the mind.

It is not a prescribed industrial disease.

55. It has been suggested that "eggshell personalities", vulnerable individuals with compulsive or dependent personalities, who perceive themselves as victims, are more likely to develop symptoms. The key to prevention is therefore said to be persuading them that the perceived symptoms are not due to "injury".

Risk Factors

WRULD is a multi-factorial condition. Identified risk factors include: abnormal posture; repetitive motion; load handling; task invariability; duration of exposure; individual susceptibility; psychological, social and cultural factors.[56]

Posture

Awkward posture of the hand, wrist, arm, shoulder, back, neck and head has been associated with the development of upper limb disorders in keyboard operators. Working at an inappropriate height due to poor desk design, and poor technique may be relevant.[57]

Neck flexion may also occur as a result of incorrect positioning of a visual display screen, and has been associated with non-specific neck and shoulder symptoms, such as cervical syndrome. Incorrect glasses may also contribute to such symptoms.

In industrial workers, working with the hands at or above shoulder height, particularly with repeated reaching, lifting or use of the arm in flexion and abduction, may contribute to rotator cuff tendinitis.[58]

Repetitive motion

Repetitive motions and power grips of the shoulders, forearms, wrists and fingers may be associated with rotator cuff tendinitis, epicondylitis and tendinitis or tenosynovitis at the wrist.[59] Trigger digit has been associated with overuse of pistol air tools. Lateral epicondylitis has been associated with prolonged use of hammers, and repeated supination and pronation of the forearm, such as when using screwdrivers.

In keyboard operators, intensive keying may contribute to the development of tendinitis and tenosynovitis in the hand and wrist. There is no generally accepted standard for a reasonable

---

56. In some jobs only one or two of these factors may be associated with increased risk. In others, many of these factors may interact in a complex way, and identification of the separate components responsible may not be easy.
57. The introduction of non-keyboard devices such as the computer mouse has added further postural components, including radial and ulnar deviation at the wrist, and shoulder rotation.
58. Examples include shipyard welders, car assemblers and painters.
59. Examples in the industrial setting include: industrial assembly workers making cars and circuit boards; dressers and packers in the meat, fish and poultry industries.

number of keystrokes per day. However, the number of keystrokes may reflect task invariability, and the number of hours spent maintaining the same posture.

Force and mechanical stress

The application of undesirable manual force may be associated with upper limb disorders. This will rarely be an important factor among white-collar workers, where the forces required to operate machinery such as keyboards are minimal.

Among rock drillers a higher incidence of both rotator cuff tendinitis and arthritis of the acromio-clavicular joint has been attributed to a combination of posture, load handling and hand-arm vibration. Repeated forceful flexion and extension movements of the wrist, particularly where associated with vibration, have been associated with CTS.

Task invariability

Many ergonomists now try to design jobs that offer physical and psychological variability of job content. Tasks that have poor content and invariability, are thought to contribute to upper limb disorders through prolonged loading of the muscles of the neck and upper limb. In industrial workers this will usually be dynamic loading, and in keyboard operators, static loading.

In the industrial setting, prevention will require job rotation and adequate rest breaks. Among white-collar workers, naturally occurring breaks which permit change of task and posture may be more important. These may include activities such as answering the telephone, filing documents and photocopying.

Duration of exposure

There will normally be a close temporal relationship between the onset of symptoms and some significant change in the system of work. There is no latent period, although there may be a cumulative effect over weeks or months. It may be difficult to prove causation in the absence of such a relationship.

Environmental factors

General environmental factors may contribute, for example poor lighting contributing to abnormal postures, and flickering lights and glare causing visual difficulties.

Psychological and social factors

Some studies suggest that perception of pain is associated with psychological and social factors. Piecework appears to be associated with a higher incidence of upper limb disorders compared with work paid on an hourly or daily basis. This may reflect the increased pace of work, tension, anxiety, boredom, or lack of control over the working environment, leading to a sense of alienation.

Similar factors may arise where increased targets for particular tasks are imposed, where there is continuous "blind" electronic monitoring of output or presence at a workstation, and where the management style is unsupportive. This may be compounded by high turnover of staff and frequent absences for sick leave.

Individual susceptibility

Some individuals may be at increased vulnerability to upper limb disorders through factors such as age, pre-existing disease such as arthritis, genetic factors, or lack of physical fitness contributing to poor posture.

Prevention

Advice and training

Employees need appropriate advice and training with regard to risk awareness and safer working practices. An induction period should permit graded introduction of work at a slower pace than the established work force.

Posture

Improved posture may require relocation of the worker, or the object being worked upon. The heights and widths of work surfaces, or particular tasks, can be redesigned to prevent reaching, raising of arms above shoulder height, or prolonged work with the elbows raised and extended.

In the case of VDU operators, detailed guidance has been provided in relation to the appropriate design of screens, desks, chairs, document holders and foot rests, and the requirement for adequate rest breaks. For example, positioning a VDU at an appropriate height will prevent prolonged neck flexion.

Repetitiveness and task invariability

Repetitiveness may be reduced by greater use of mechanical aids, automation of certain tasks, enlargement of job content through greater diversity of tasks, job rotation, and adequate rest periods. For example, automation of food packing tasks has been shown to reduce the incidence of neck, elbow and wrist symptoms.

Force and mechanical stress

The mechanical stresses from hand tools can be reduced by ensuring optimal handle size, shape and friction, so that force is distributed over the maximum possible area of the hand, thereby avoiding palm pressure. Knives have been altered to reduce grip force and forearm fatigue in poultry dressers, and hammers altered to reduce ulnar deviation in carpentry.

Excess loading can be reduced through decreased weight of tools, containers, and the parts being worked upon. It can also be reduced through improved mechanical advantage and torque control devices.

Psychological and social factors

The impact of these factors can be reduced by enlargement of job content, more individual control over the pattern of work, short informal work pauses, reduction of paced work and continuous electronic monitoring.

Treatment

Many cases of WRULD will resolve with conservative treatment such as rest and non-steroidal anti-inflammatory drugs (NSAID's). Rest from the task implicated may require temporary restrictions of duty, job rotation or job transfer. Complete removal from the working environment is usually only reserved for the most severe cases.

Other physical therapies which may be effective include heat, massage, ultrasound and physiotherapy directed to stretching and strengthening of muscles. In persistent cases local steroid injections may be effective.

Splints or supports should be used with care, and only where medically indicated to avoid bending or deviating the affected joint. Inappropriate usage may result in muscle atrophy through disuse, and increased stiffness.

As a last resort, surgery may be indicated, for example to release trigger digits, to decompress the carpal tunnel in CTS by division of the transverse carpal ligament, and to remove the lateral acromion in chronic severe shoulder tendinitis.

When the symptoms have subsided, it should be possible to gradually reintroduce normal duties, having ensured that the necessary changes to the working environment have been addressed, such as inadequate workstation ergonomics, poor technique and task invariability.

Further Reading

Palmer K, Cooper C. "Repeated Movements and Repeated Trauma Affecting the Musculo-skeletal System" in: *Hunter's Diseases of Occupations* (9th edition), London, Hodder, 2000.

Hagberk, M. "Neck and Arm Disorders". In: *ABC of Work Related Disorders*. London, BMJ Publishing Group, 1997.

CHAPTER 8
NOISE INDUCED INDUSTRIAL DEAFNESS

SECTION 1: THE LEGAL PERSPECTIVE

Overview

In a claim based on hearing loss induced by noise at work, a claimant will have little difficulty in showing that he was owed a duty of care by reason of the employer/employee relationship. The more complex issues relate to the breach of the duty of care and causation which are critical to the successful prosecution or defence of a claim. There will often be complex points on causation, for example whether and to what extent a hearing loss has occurred and whether the cause can be ascribed to the defendant employer's system of work, or part of the normal aging process.

Foreseeability of Injury

The Knowledge of the Employer of the Risk of Injury

That excessive noise at work can cause deafness has been widely known for a number of years. Accordingly with causes of action based on injury from the 1980s onwards it will be difficult for an employer to claim that it did not know (actually or constructively) that employees were at risk of injury from excessive noise exposure at work. Proof of the date of knowledge is supported by the following:

(a) Noise and the Worker: [Safety, Health and Welfare Booklet No.25] 1963[1]

1. Note that in *Thompson* v *Smiths Shiprepairers Limited* [1984] QB 405 the relevant date of knowledge of the employer was set at 1963.

(b) "Code of Practice for Reducing the Exposure of Employed Persons to Noise" [Dept of Employment: 1972].

(c) Social Security (Industrial Injuries) (Prescribed Diseases) Regulations 1975.[2]

(d) Noise at Work Regulations 1989.

(e) Provision and Use of Work Equipment Regulations 1992 and 1998

(f) Personal Protective Equipment at Work Regulations 1999.

Breach of Duty

Breach of Duty at Common Law

The appropriate point of entry is first to establish that a claimant has suffered a hearing loss caused by noise at work. If causation is not proved then the claim will fail. This issue is dealt with in the section on causation. But assuming that causation is proved, a claimant still has to discharge the burden of proof that the employer breached its duty of care.

In general, an employer is under a duty at common law to take reasonable steps to prevent an employee suffering noise induced deafness at work. The standard of care is that of the reasonably competent employer.[3] If an employer follows an accepted practice which accords with a sensibly held view of what is safe then the defendant may not be in breach of its duty.

The proper approach to breach of duty was set out in *Thompson* v *Smiths Shiprepairers (North Shields) Limited*.[4] The claimant worked in the defendant's shipyard for a number of years. The defendant knew there was a risk of deafness, but it only provided ear protectors from the early 1970s. The claimant suffered a loss of hearing from prior to the 1960s although his condition deteriorated thereafter. First, it was held that knowledge of the risk of injury was in or about 1963. Secondly, once the risk of injury was known in the industry then the duty meant that ear protectors should have been provided.[5]

---

2. Disablement benefit is paid to persons who suffer at least 50 dB of noise induced hearing loss.
3. *Stokes* v *Guest et al* [1968] 1 WLR 1776.
4. [1984] 2 All ER 881/[1984] QB 405.
5. See also *Berry* v *Stone Mangenese Marine Limited* [1972] 1 Lloyd's Rep 182.

Breach of duty depended in part on when the noise exposure became so excessive as to be negligent. It was held that:

(a) the test is based on the reasonable and prudent employer who was "properly" solicitous for its employees' safety.

(b) Following a recognised practice could be acceptable even if it presented certain risks of injury, although this depends on the circumstances of each case.[6]

> "Between the two extremes is a type of risk which is regarded at any given time (although not necessarily later) as an inescapable feature of the industry. The employer is not liable for the consequences of such risks, although subsequent changes in social awareness or improvements in knowledge and technology may transfer the risk into a category of those against which the employer can and should take care."[7]

(c) A failure to provide ear protection before 1963 would not be negligent, but thereafter such an omission would be negligent.[8] The nature and extent of hearing protection will depend in part on a proper risk assessment being carried out and implemented.

Statutory Duty

The statutory regulations relating to noise induced hearing loss include the following:

(1) *Factories Act 1961, section 29(1)*:[9] imposed a duty on the employer to make and keep the place of work in a safe condition.

(2) The *Noise at Work Regulations 1989*:

- *Regulation 2(1):* an employer must make a formal noise assessment above 85dB.

---

6. See also *Morris* v *West Hartlepool Steam Navigation Co Limited* [1956] 1 All ER 385.
7. Mustill J at 889: the issue of whether technology has changed the duty of care is a point to be covered by expert evidence.
8. For the need for ear protection in the case of firearms see *McCafferty* v *Metropolitan Police Receiver* [1977] 2 All ER 756. [1977] 1 All ER 756.
9. Now repealed.

- *Regulation 4:* the assessment has to be performed by a competent person and there is an on-going duty to review the situation.
- *Regulation 6:* employers have to reduce the level of noise to the lowest level reasonably practicable.
- *Regulation 7:* employers must take all reasonably practicable steps to reduce the exposure of employees to noise where the level is 90dB or above.
- *Regulation 8:* where an employee will be exposed to noise at 85dB or above then suitable and efficient ear protection must be provided by the employer.
- *Regulation 9:* designated ear protection zones must be set up and ear protection must be worn in these areas.
- *Regulation 10:* all ear protection must be maintained in an efficient state and the employer must ensure it is properly used.
- *Regulation 11:* employers must warn employees of the risk of damage to their hearing and how to minimise that risk, in addition to instruction and training, where the noise exposure is greater than 85dB.

Personal Protective Equipment at Work Regulations 1992 (as amended by the 1999 regulations):

- *Regulation 6:* requires an assessment of the provision and suitability of personal protective equipment.
- *Regulation 4:* requires the provision of suitable personal protective equipment.
- *Regulation 9:* the duty is on the employer to ensure that protective equipment is used.

Contributory Negligence

The court's approach to the issue of contributory negligence is illustrated by *McCafferty* v *Metropolitan Police Commissioner.*[10]

10. [*Supra*] at note 8.

The claimant was a liaison officer working with firearms in the laboratory at Scotland Yard. He worked under the supervision of scientists and he became a leading expert on ballistics. Ultimately, he suffered a loss of hearing. The claimant was initially led to believe by the other scientists in the laboratory that cotton wool stuffed in his ears was sufficient to provide adequate protection. Later he used ear muffs, but was not aware of the need for structural acoustic protection.

The defendant contended that the claimant had contributed to his injury in that authority was delegated to him to make enquiries about safety which he had failed to do. Specifically, he should have known that cotton wool was a useless form of protection and if he had discharged his authority properly and informed the defendant then his injury would have been less serious. This contention was rejected by the Court of Appeal. The claimant could not be blamed for not realising that cotton wool in his ears would not be sufficient protection, as he was led to believe this by other scientists at the laboratory. Moreover he lacked the necessary knowledge about his hearing protection and he was required and entitled to believe that ear muffs would provide sufficient protection.

Causation

A Summary of the Main Principles

Causation is considered in detail in chapter 2, but in summary the following points ought to be noted:

(a) A claimant has to prove causation, namely that the negligence or breach of statutory duty caused the injury for which damages are claimed.

(b) Causation in industrial deafness cases will have to take into account non-tortious and other contributory factors. The liability of the defendant is based only on the extent to which the negligence, or breach of statutory duty contributed to the injury.

(c) Accordingly, expert evidence will have to cover all causal factors:

- The pre-existing level of hearing.
- The loss of hearing induced by natural causes.
- Hearing loss caused by non-negligent exposure in the workplace (e.g. some loss of hearing is bound to be caused by working in a noisy environment).
- Hearing loss caused by previous employers.
- Hearing loss caused when no duty of care was imposed on the defendant (i.e. before the risk of injury was known).

An Example of the Approach of the Courts to Causation

In *Thompson* v *Smiths Shiprepairers (North Shields) Limited*[11] causation and apportionment was approached on the following basis:

> "I reject the submission that *McGhee*[12] demands a recovery for the full amount of impairment in the present case. Any other conclusion would accord ill with the real shape of the issues....I cannot think it legitimate for a plaintiff to assert an evidential gap and then use the *McGhee* principle to bridge it by a device that enables him to recover in full for a loss, part of which is known to have occurred before the wrongful conduct began."[13]

Accordingly, a claimant could not claim for all his hearing loss merely by proving that the defendant materially increased the risk of injury through excessive noise. In addition, the failure by the employer to take regular audiograms did not entitle claimants to claim in full for the injury caused. Having rejected that contention Mustill J. acknowledged the difficulties in quantification of damages in this type of case, but sought to apportion damages accord to the contribution of the defendant to the injury.

11. [*Supra*] at note 4.
12. *McGhee* v *NCB* [1973] 1 WLR 1 HL.
13. Mustill J at 909.

He approached the issue of contribution as follows:

> "I see no reason why the present impossibility of making a precise apportionment of impairment and disability in terms of time should in justice lead to the result that the defendants are adjudged liable to pay in full when it is known that only part of the damage was their fault. What justice does demand... is that the court should make the best estimate which it can in the light of the evidence making the fullest allowances in favour of the plaintiffs for the uncertainties known to be involved in apportionment."[14]

Limitation

Recent Case Law

Progressive deafness frequently raises limitation issues. Limitation is dealt with in detail in chapters 4 and 5, but a recent example that demonstrates the proper approach to deafness cases is *Ali* v *Courthaulds Textiles Limited*.[15] The claimant worked for about twenty years in a cotton mill that was operated by the defendant. In 1990 he complained to his general practitioner that his hearing was deteriorating. At the end of 1990 he was informed by a community worker that this could have been work related. The claimant went to a solicitor and an ENT consultant reported in 1992 that his hearing was adversely affected and that it was probably caused by the noise at this factory.

Proceedings were commenced in 1995 against the defendant alleging an unsafe system of work and the defendant relied on a limitation defence. Limitation was tried as a preliminary issue and the claim was dismissed. The Court of Appeal allowed the claimant's appeal and held that the claim was brought within the limitation period. It made the following findings:

(1) The conversation with the community worker in 1990/91 when the claimant was advised about possible noise induced hearing loss did start the limitation period. But there was a total failure to consider section 14(3) of the Limitation Act 1980 which highlighted the need to obtain medical evidence to establish knowledge for the purposes of limitation.

---

14. Mustill J at 910.
15. [1999] Med LR 301. See also *Milner* v *Hepworth Heating Ltd* [1998] LTL 28/7/98.

(2) The claimant knew he was going deaf, but was unclear if it was due to the ageing process. That could only be discovered by expert advice.

(3) Relevant knowledge which can only be gained through expert medical evidence is qualified in that it is knowledge "he might reasonably be expected to acquire" from his medical expert. Appropriate steps therefore had to be taken to obtain the medical evidence within a reasonable time. Whether an individual claimant's knowledge of the cause of injury requires medical expert evidence will depend on the facts.

Quantum

Set out below in the schedule is a series of recent cases that provides an indication of the level and range of damages for noise induced hearing loss injury:

Case Name	Comment
Rooney v *Guinness Plc* [1999] S.L.T. 700	R, a female factory worker aged 58, allegedly suffered deafness at work. She had difficulty in hearing conversations. She required the volume of her television to be turned up and a louder telephone. She tended to avoid social situations involving crowds of people. *GD: £3,000*
Earlam v *Hepworth Heating Ltd* [1997] Current Law 1878	Male, aged 46, suffered significant hearing loss for 4–5 years attributable to exposure to loud noise at work. He suffered no tinnitus. The television was turned up too loud and he could not follow conversations. Difficulty in hearing the telephone. Otherwise E's handicap did not interfere with any hobbies. Loss: at 1 kHz: right ear 10dB; at 2 kHz: right ear 20 dB and left ear 15 dB; at 3 kHz right ear 45dB: 50% chance he required a hearing aid sooner than a person not exposed to noise. *GD: £5,500*

Case Name	Comment
Lewis v *BTR Plc* [2000] 1 Q.R. 7	L, male, 35 suffered high frequency sensorineural hearing loss in both ears as a result of exposure to loud noise at work. He could follow conversations without background noise. He had problems conversing whilst in a group and when being spoken to on shopping trips. Television was at high volume. He did not complain of tinnitus. Loss: left ear 10dB at 1kHz, 15dB at 2kHz and 15dB at 3kHz. Right ear 20dB at 1kHz, 15dB at 2kHz and 15dB at 3kHz. L's binaural hearing loss averaged over 1, 2 and 3kHz was 13.33 dB. *GD: £5,500*
Owens v *Express Group* [1991] QBD Drake J: Lawtel 31.1.91	Male 56, exposed to high noise levels at work. Suffered a loss of hearing with tinnitus disturbing his rest. Difficulties in hearing music and more than one person talking. *GD: £10,125, after 25% reduction for injury caused whilst working for other employers.*
Holland v *Hoechst Trespaphan* [2001] Lawtel 12.3.01	Male claimant 52, was exposed to excessive noise between ages 24 and 48. Suffered severe bilateral hearing damage. Average hearing loss at 1, 2, 3kHz of 30 decibels in left ear and 32 decibels in right ear. He also suffered tinnitus which was permanent and constant. Occasional headaches. Attempted to use NHS hearing aids. Injuries were permanent and would deteriorate: *GD: £12,500. Digital hearing aids: £3340; 2 replacements: £4864; annual hearing aid maintenance cost £100 pa.*

Case Name	Comment
Hill v *ARC (South Wales) Ltd* [1998] Current Law 1563	H, male 48 exposed to excessive noise levels at work. 2 years after the onset of tinnitus, H's hearing loss was a binaural impairment over 1, 2, 3 khz of 29.33 dB. Some reduction for previous exposure but none in respect of the tinnitus. Tinnitus consisted of a ringing and a hissing sound. Some habituation was probable. Tinnitus was severe. He also suffered depression and required continuous psychiatric treatment. The chances of finding employment whilst suffering significant depression were negligible. *GD: £19,500. Future cost of hearing aid (agreed): £2,750.*

Summary of the Chapter

A summary of the main points of the chapter are set out in the schedule below, which provides practical guidance for both claimant and defendant practitioners:

Point/Issue	Claimant's Perspective	Defendant's Perspective
Knowledge of risk: foreseeability Based on the government publications and statutory regulations. The date of knowledge of the risk of injury is about 1963 onwards. But a possible duty of care before that date can be argued.	• Reliance placed upon the publications outlined in the chapter. Foreseeability of injury will probably not be a problem in most cases.	• Consider carefully if the date of knowledge of risk of injury can be conceded. Post 1963 it is difficult to argue successfully.
Breach of duty At common law based on the conventional common law approach. Failure to warn of the dangers, and failure to provide proper ear protection post 1963 will probably be negligent	• Obtain detailed statement from claimant, indicating full CV and all employers. Main points will be warnings and instruction on ear protection and failure to provide ear protection. • The precise nature and extent of exposure needs to be clarified. • Consider the risks of injury from noise in the employment in question and assess the magnitude of risk. • When did the claimant tell the employer of his problem?	• Obtain a statement from the claimant so that the complete version of events is known early on in the case. • Witness evidence from line managers and colleagues on the extent of noise exposure on the floor. Directors also have to indicate all health and safety compliance. • What is the extent of the risk inherent in the job? • Pin the claimant down on the version of events as soon as possible. Part 18 requests if necessary. • Investigate other employers and consider Part 20 proceedings at an early stage.

Point/Issue	Claimant's Perspective	Defendant's Perspective
Breach of statutory duty Main regulations are Noise at Work Regs 1989: Other applicable statutory regulations (Workplace (Health, Safety and Welfare) Regulations 1992, Personal Protective Equipment at Work Regulations 1992, 1999). Provision and Use of Work Equipment 1992/98.	• Expert evidence is needed on statutory duty and causation. Based on a detailed statement from claimant. Periods and intensity of exposure are vital. • Consider the extent to which risk assessments have been carried out. • Consider compliance with the regulations. All health and safety documentation on policy and implementation needs to be disclosed. • Consider contributory negligence.	• Obtain expert evidence based on the regulations. All health and safety documentation showing reasonable compliance with regulations must be considered. • Does the case fail on causation? Are any breaches actually causative of the injury alleged? • Evidence from witnesses with experience of working in this environment. What is the situation with other colleagues? • On the medical notes when did the symptoms first present? • Is there an arguable case of contributory negligence?

Point/Issue	Claimant's Perspective	Defendant's Perspective
Limitation Knowledge of injury under s.14(1) of the Limitation Act 1980 is important. In many cases expert evidence is required to clarify causation. If the claim is time barred then a swift application under s.33 is required.	• Clarify when the claimant first made the connection between the injury and the work. • Obtain all DSS/Benefits' Agency and medical records and ensure there are no indications of earlier knowledge of injury. Do the presentation of the symptoms tie in with the noise at work? • What is said about previous employment?	• Obtain all medical and DSS/Benefits' Agency records and assess if there is an earlier date of knowledge. • Can it be shown that the claimant knew about the connection between the injury and work many years before the date alleged?
Causation Medical evidence must prove there is a loss of hearing and that it was caused by the noise at work. If that is proved then the assessment is based on the contribution of the defendant to the injury. There is no right to full compensation in the absence of full attributability.	• Obtain a medical report to cover causation. The medical evidence must detail the hearing loss and if necessary attribute competing causes. • Causes can be both previous employers and/or other non-tortious factors.	• Obtain medical evidence. This must cover the competing causal factors. • First, whether the system of work could have caused this particular hearing loss. Secondly what are the other competing causal factors that reduce the contribution of the defendant.

SECTION 2: THE MEDICAL PERSPECTIVE

Overview

Deafness may arise through disease or injury to the sound conducting system of the ear (conductive deafness), or to the system which conducts nervous impulses to the auditory centres of the brain (perceptive or sensori-neural deafness). Noise induced hearing loss ("NIHL") is an example of sensori-neural deafness. However, the commonest cause of sensori-neural deafness in adults is age-related degeneration ("presbyacusis").

The association between occupational noise and hearing loss has been known for many hundreds of years.[16] One study performed in the 1980's found that 12% of all British adults had sensori-neural hearing defects, of which about one third were due to age, and 5% due to noise exposure.

Basic Anatomy (see Figure 5)

Sound waves are collected by the trumpet-shaped outer ear (the "auricle" or "pinna"), and travel through a narrow bony canal, the external auditory meatus, where they reach a thin elastic membrane, the ear drum (the "tympanic membrane").

16. It was described by Ramazzini in copper workers in the early eighteenth century, and by Barr in Glasgow boilermakers in the mid-nineteenth century.

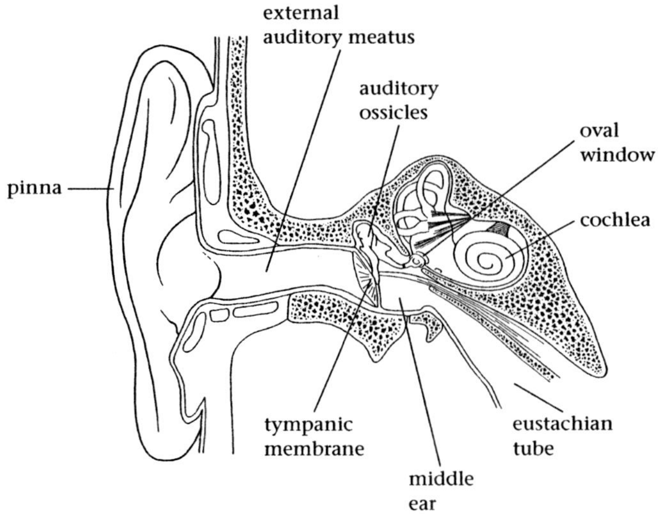

Figure 5

The middle ear lies beyond the ear drum. It is an air-filled space which is connected to the throat at the back of the nose by the Eustachian tube.[17] The middle ear also contains three tiny bones, the auditory ossicles.[18]

The hammer is attached to the inner surface of the ear drum, and transmits sound vibrations to the stirrup via the anvil. The stirrup is in turn attached to the oval window, another thin elastic membrane separating the middle ear from the inner ear.

Within the inner ear, vibration of the oval window stimulates a tiny fluid-filled spiral structure shaped like a snail (the cochlea), within which is the organ of hearing (the organ of Corti). This structure contains thousands of hair-like nerve endings, the stereo-cilia, which convert acoustic mechanical energy into electrical energy.

These electrical impulses are then conducted along the fibres of the auditory nerve (the eighth cranial nerve), which pass through another narrow bony canal, the internal auditory

17. The Eustachian tube permits equalization of pressure when ascending or descending through air or water.
18. Known individually as the hammer (malleus), the anvil (incus) and the stirrup (stapes).

meatus, before reaching the brainstem, then finally the auditory cortex of the brain.

Basic Physics

The application of force to an elastic medium such as air sets up a vibration or oscillation of the molecules forming that medium, in other words, a sound wave. This can be plotted as a graph showing pressure change as the vertical axis, and time as the horizontal axis.

The basic components of the sound wave are:

(a) *Intensity*. This is the peak height of the pressure wave, which corresponds to loudness.

(b) *Frequency*. This is the number of completed cycles of sound wave per second, which corresponds to pitch. One cycle per second is described as one "Hertz" (Hz).

Intensity

Intensity is defined as the amount of acoustic power transmitted by the sound wave over a standard unit of area. This is measured in Watts[19] per square metre. Unfortunately intensity, or transmitted acoustic power, cannot be easily measured.

However, pressure can be easily measured, using a microphone. This is expressed as the force transmitted by the acoustic pressure wave over a standard unit of area. This is measured in Newtons[20] per square metre, otherwise known as "Pascals"[21] (Pa). Sound intensity varies as the square of sound pressure. This means that doubling the pressure will increase the intensity by $(2)^2$ i.e. a factor of four.

However, the pressure of a sound wave does not remain constant. The wave is cyclical, so that if pressure is measured at any instant, it may be at a peak or a trough value. Thus in order to calculate the average pressure transmitted by the wave, a value known as the "root mean pressure" (RMS) is used to provide an average.

19. The basic unit of power.
20. The basic unit of pressure.
21. Thus a force of one Newton applied to one square metre is 1Pa.

However, there are a number of practical difficulties with the use of unmodified pressure readings:

(a) The human ear is capable of detecting an enormous range of pressures.

(b) The ear does not respond in a direct or linear fashion to linear changes in sound pressure.

This is overcome by applying an arithmetical technique known as "logarithms" to the pressure readings. First, the measured RMS sound pressure is compared to an international standard reference pressure. The logarithm of this ratio is then calculated, and the figure derived for the sound pressure level is defined as a "bel".[22] In fact one tenth of a bel is more convenient to use, the "decibel" (dB).

Using the decibel scale for "sound pressure levels" (SPL), 0 dB SPL is the pressure threshold at which faint sounds are first heard. A sound pressure level of 130 dB SPL, is the threshold at which the sound is so loud, it will seem painful.

A simple arithmetical formula can be used to calculate the sound intensity, or acoustic power transmitted, from the sound pressure level. The sound intensity is described in the same units as sound pressure, in other words dB SPL.

A change of 10 dB SPL in sound intensity means that the intensity has increased by a factor of ten. However, because the ear does not respond in a direct or linear way to either sound pressure or power, this will be perceived subjectively as a mere doubling of "loudness".

Hearing threshold level

There is considerable individual variation in the ability to detect faint sounds. The level at which sound is first perceived is described as the hearing "threshold". This can be measured using a device known as an audiometer. For example, an individual might be unable to hear any sound until the reference sound pressure level reaches 20 dB.

---

22. Named after Alexander Graham Bell.

Using the reference pressures described above, an audiometric reference sound level of 20 dB would be described as 20 dB hearing level (20 dB HL). If this is the same as the hearing threshold for that ear, this is described as 20dB hearing threshold level (20 dB HTL).

The effect of frequency

To further complicate matters, the threshold at which the human ear can detect sound also varies according to the frequency or pitch. A doubling of frequency will be perceived as a change in pitch of one octave. Middle C has a frequency of about 250 Hz. Tuning fork and audiometric tests of hearing usually use multiples of middle C, from 125 Hz to 8 kHz.

In a child the range of frequencies which can be detected may vary from as low as 20 Hz up to 20,000 Hz (20kHz). The ability to detect these higher frequencies is lost with normal aging. Very few adults are able to detect frequencies higher than about 15kHz after the third decade.

Most important speech sounds have frequencies in the range 2kHz to 3kHz. For these reasons, when assessing sound pressure levels or hearing thresholds, the very low and very high frequencies are corrected by weighting them less heavily than the middle frequencies. The "A" weighting scale is most commonly used. When this is done, the sound pressure level appears with the prefix "A", e.g. 20 dB(A) SPL.

Measuring noise in the industrial setting

"Noise" can be classified as continuous or intermittent. Intermittent noise can be further classified as "impulse noise", and "impact noise". A typical example of "impulse noise" is that due to gunfire. The noise is of short duration, but has a significant high-pressure shock wave. "Impact noise", such as that caused by hammering or riveting, has an insignificant pressure shock wave, but more significant reverberating sound, emanating from the machinery.

Owing to noise fluctuations in the occupational setting, integrating sound level meters have been developed. They can calculate a single average "A" weighted sound level, with the same energy content as the sounds being measured. This is the "equivalent sound level" (L_{eq}). When taken over an eight-hour shift, or reduced to the equivalent of eight hours, it is referred to as the L_{EPd}. This is the daily personal exposure to noise, the personal noise "dose" received.

A daily personal noise exposure of 85 dB(A) is defined as the "first action level", requiring ear protection to be available. 90 dB(A) is the "second action level", where ear protection is mandatory. Ear protection is also mandatory if any single instantaneous peak pressure exceeds 200 Pa, regardless of the overall noise dose received.

Basic Pathophysiology

Temporary threshold shift

The human ear is able to respond to sound pressure levels of up to 80dB by an immediate elevation of hearing threshold. This is automatically reset within a few seconds. This normal response is known as "adaptation" or "pre-stimulatory fatigue". At higher sound pressure levels, an abnormal "temporary threshold shift" or "post-stimulatory fatigue" will occur.[23]

The degree of temporary threshold shift depends upon the intensity and duration of the sound stimulus. There is considerable individual variation, and the frequency of the sound is also relevant. In practice this means that recovery may vary both according to the individual and the frequency of the sound. Recovery periods vary, but may take several days, particularly with higher frequencies.

Depending upon the noise level, and the other factors set out above, these temporary threshold shifts may become permanent. Typically, the rise in threshold is at higher frequencies, usually between 3kHz to 6kHz and centred on about 4kHz. This appears as a dip or notch in the "audiogram", the graph showing hearing thresholds measured by audiometry across a range of frequencies from 125 Hz to 8 kHz (see figure 6).

23. Often there is associated tinnitus, apparent noise in the ears where there is no acoustic source. This may be perceived as buzzing, ringing or a growling noise.

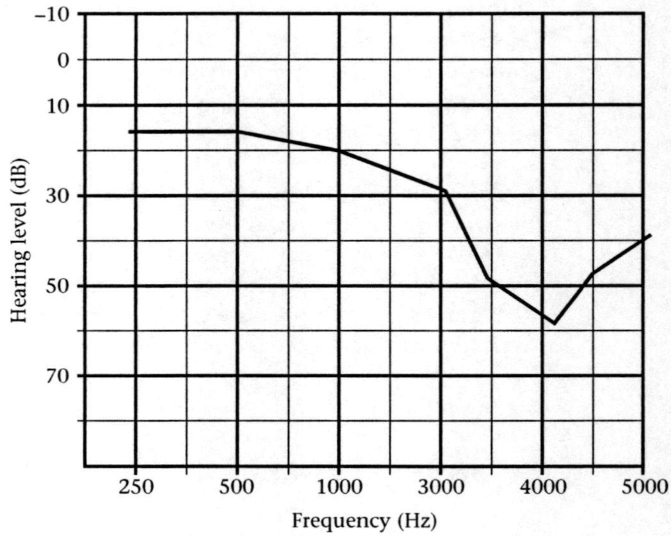

Figure 6

With further exposure, the notch gradually deepens and widens, to involve the speech frequencies in the range 2–3kHz. Once this occurs, the affected individual may notice:

(a) Diminished auditory acuity.

(b) Difficulty in discriminating speech from other background noise.

(c) Tinnitus, particularly after prolonged exposure to impact noise.

The mechanism is not clearly understood, however, the end result is progressive damage to and then death and loss of the stereo-cilia and their supporting hair cells in the organ of Corti.

Diagnosing NIHL

Clinical history

NIHL commences with a gradual loss of clarity of perceived speech, particularly where there is background noise (perceptual rivalry). Some sufferers notice sound suddenly becoming more intense at a certain volume (loudness recruitment). There may also be tinnitus, varying both in frequency and intensity throughout the day. In some cases this is more disabling than the actual hearing loss.

A thorough assessment of the employment history, social history and past medical history is essential. The main points are:

(a) Details of all employers since leaving school or college.

(b) The number of years with each employer

(c) The nature of the duties.

(d) The degree of exposure to noisy machinery.

(e) The nature of the available ear protection, such as ear plugs or muffs.

(f) The periods when ear protection was actually used.

(g) Where relevant a detailed history of exposure to occupational and recreational gunfire,[24] including the nature of the weapons, the approximate number of rounds fired, what ear protection was available, when it was worn, and which side the weapon was fired on.

(h) Details of exposure to loud music at concerts, clubs or through other amplification systems.

There are numerous causes of deafness other than NIHL, of which age-related changes and conductive deafness due to recurrent middle ear disease are the most common. However, sometimes the cause of deafness is unknown. The past medical

24. The forward ear, that is to say the left ear in a right-handed individual, is usually more badly affected by pistol or rifle fire. It is also important to note that a single discharge, which may peak at up to 190dB for a field gun, may be sufficient to cause NIHL.

history will exclude many less common causes of sensorineural or conductive deafness, for example:

(a) Exposure in the womb to infections such as German measles (rubella) or certain drugs.

(b) Birth trauma.

(c) Childhood illnesses (measles, mumps, meningitis, encephalitis).

(d) Basal skull fracture.

(e) Ménière's disease.[25]

(f) Excess exposure to or overdose of certain ototoxic drugs (streptomycin, neomycin, gentamicin, quinine, aspirin).

(g) Otosclerosis.[26]

(h) Paget's disease.[27]

Clinical examination

Examination requires visual inspection of the external auditory canal and eardrum by means of an otoscope.[28] This might reveal evidence of old disease of the external or middle ear, such as recurrent infections of the middle ear (otitis media), with resultant scarring and retraction of the eardrum.

Diagnostic tests

The principal diagnostic investigation is pure tone audiometry. The patient sits in a sound proof room, wearing headphones. The threshold at which they respond to a range of pure tones is plotted on a graph, the audiogram. It is usual to obtain estimates of average hearing loss at 1, 2 and 3 kHz, and a binaural average.

In cases of suspected conductive deafness, for example due to middle ear disease, it is sometimes necessary to plot separate

25. A disease of the inner ear which results in disabling attacks of tinnitus and vertigo.
26. A degenerative condition of the auditory ossicles, which become fused.
27. An inflammatory condition of bone, which may effect hearing, for example by narrowing of the auditory canals.
28. This is usually supplemented by tuning fork tests, used to screen for differences between bone and air conduction, and asymmetrical hearing loss.

graphs for air conduction and bone conduction. The latter is performed using a bone vibrator placed over the mastoid bone, behind the ear.

A significant difference between the air and bone conduction levels suggests that conductive deafness is the cause of the hearing loss. However, bone conduction is inaccurate above 3 kHz.

Practical interpretation of the audiogram

The audiogram pattern is not diagnostic in itself.[29] For example:

(a) The characteristic 4kHz notch may occur in individuals who have never been exposed to significant noise.

(b) With increasing loss of thresholds at other frequencies, the notch may be lost completely.

(c) The audiogram changes are not necessarily symmetrical, for example when the leading ear is more badly affected by gunfire. In the absence of such an obvious cause, asymmetrical hearing loss may be difficult to assess.

(d) There may also be overlapping effects due to normal aging.

Assessment of Hearing Disability

The low fence

The "low fence" is the point at which elevation of the hearing threshold, measured by audiometry, is deemed to be first disabling. As indicated below, the low fence for the statutory compensation scheme is deemed to be 50 dB HL. In reality, the handicap at this level is "substantial".

However, arguably, the true disability threshold or low fence is nearer 26 to 30 dB HL. If the impairment is about the same for

29. Where there is uncertainty it is sometimes necessary to perform other tests. Tympanometry measures the sound energy reflected from the ear drum, and is useful in detecting stiffness in a previously diseased drum. Electrical response audiometry (evoked response audiometry) studies the response of the auditory pathways to sound, by mapping the electrical activity from the cochlea to the brain using electrodes placed on the head. The latter is sometimes used to detect suspected exaggeration of hearing loss, since the subject cannot influence the results. Other tests which are occasionally employed include: auditory brainstem responses (ABR) and electro-cochleography (EcochG).

both ears and there is no tinnitus, it is unusual for the sufferer to experience significant disability with less than 20 dB HL.

The high fence

The "high fence" is the point at which disability is deemed to be total. Again, arbitrary levels have been used, at any level from 70 dB HL upwards.

Adjustment for age-related changes

In assessing hearing disability, adjustments need to be made for age-related changes, which are published in standard tables. The inter-relationship between NIHL and age-related changes is complex. However, the age-related changes progressively accelerate with time, and the contribution of noise decreases.[30]

NIHL is a prescribed industrial disease (A10) where there is "substantial" sensori-neural hearing loss. This is defined as at least 50 dB hearing loss in each ear, using the average loss over 1, 2 and 3 kHz, where in the case of at least one ear, the loss is due to occupational noise.

Prevention

Noise reduction

Many techniques have been used to reduce noise emissions, including silencers, mufflers, damping, anti-vibration mountings, and the use of sound absorbing materials. Manufacturers of equipment are required to provide noise emission levels.

Protective equipment

Personal protective equipment has an important role to play.

Ear muffs[31] should fit over the ears and be sealed to the side of the head by foam or enclosed fluid seals. Some are frequency selective, and protect at higher frequencies more than lower frequencies. Some are amplitude selective, and offer increasing resistance as the sound pressures rises.

30. Thus by the age of 80, for example, the relative contribution of noise is negligible.
31. Ear plugs are easier to use, but have the disadvantage that they are less efficient than ear muffs, and have a shorter working life.

The sound attenuation should be to about 80dB, beyond which social isolation may be a problem. Manufacturers usually provide the mean and standard deviation of the ability to attenuate standard frequencies or a single number rating (NNR). The level of protection is conventionally considered to be the mean, less one standard deviation, per frequency.

Ear muffs have the disadvantage that any breach of the seal, for example due to spectacles, beards, or long hair, will greatly decrease their effectiveness, as will wear and tear. Furthermore, removal, even for short periods, may have a significant impact on the amount of protection provided.

Examples of protection in practice

Exposure to 100 dB(A) for only 15 minutes is equivalent to eight hours exposure at 85 dB(A). The best available protectors worn for only half the time will only reduce the L_{EPd} by 3–5 dB(A). In conditions of extremely high noise muffs or plugs may not be adequate, and noise-excluding helmets have been developed. The provision of "ear protection zones" is also common.

Workers exposed to a risk of NIHL need an adequate programme of education, training and information, including information about the warning symptoms of temporary threshold shift and tinnitus. Pre-employment audiometry should be followed by screening audiometry at regular intervals.

Treatment

Apart from the phenomenon of temporary threshold shift, NIHL is permanent, and there is no treatment apart from hearing aids. The focus is therefore on preventing further damage. This usually requires removal of the individual from the noisy working environment, or complete protection from the noise.

Further Reading

Ramsden, R and Saeed, S. Sound, "Noise and the Ear" in: *Hunter's Diseases of Occupations* (9th edition). London, Edward Arnold, 2000.

Jones, C. "Occupational Hearing Loss and Vibration Induced Disorders". In: *ABC of Work Related Disorders*. London, BMJ Publishing Group, 1997.

CHAPTER 9
HAND-ARM VIBRATION SYNDROME

SECTION 1: THE LEGAL PERSPECTIVE

Overview

Hand-Arm Vibration Syndrome ("HAVS" formerly referred to as "Vibration White Finger") is an injury that affects the hands, fingers and the upper limbs. The vibration can cause injury to the vascular, nervous and musculo-skeletal systems but it covers all types of injury to the hands and lower arms. The detailed particulars about the condition are set out in the medical section. The majority of claimants have been using powerful vibrating hand held tools, although claims are not limited to this type of cause (see, for example, injury driving cranes). The main issues to address in bringing and defending these claims are:

(1) Breach of duty: specifically what was the duty of care, the system of work and the extent of the use of the tools in question.

(2) Foreseeability: when the employer should first have known about the potential harmful effects of using a particular type of tool and whether the injury suffered was reasonably foreseeable.

(3) Causation: whether the injury in question was caused by the use of the tools in question.

Many such claims have been settled under a mediation scheme. For example, a series of claims was brought against British Coal Corporation and these have been consolidated and mediated to settlement.[1] The scheme has a deadline of 31st of January 2003 in relation to deceased miners and the deadline for live claimants

1. As to the procedure for issuing claims see Practice Note (Vibration White Finger Actions) (No.2) (QBD) Queen's Bench Division *Times*, August 5, 1999; Practice Direction (Sup Ct: Vibration White Finger Actions) (No.1) *Times*, January 13, 1994.

has now passed. Under the terms of the scheme the miners had to show they used hand held vibratory tools in excess of three hours at a time after 1975. In addition it also had to be shown they had symptoms within two years of leaving British Coal. The claims are treated like conventional personal injury claims but if they are not brought within the time period as set out above then the conventional rules on limitation will apply.

The nature and extent of the duty of care

The duty of care of the employer is based on the employer/employee relationship and where relevant, statutory regulation. The duty on the employer is actively to consider the safety of the employees, based on the knowledge that ought to have been known at the relevant time. Employers have to take reasonable steps to prevent vibrating tools from injuring employees as part of the industrial process in question. The standard of care is summarised in *Stokes* v *Guest, Keen and Nettleford (Nuts and Bolts) Limited*.[2] A detailed assessment of the duty of care is contained in Chapter 1.

Foreseeability:

Knowledge of the Risk of Injury

An important issue is whether the exposure to vibration will give rise to a foreseeable risk of injury and if so, by what date this should have been known by the employer. The knowledge of an employer (actual or constructive) of the risks posed by vibrating tools, will fix the date at which the duty of care arose. This in turn, determines when an employer should have taken positive steps to protect employees from the possibility of injury from vibrating tools. The date of knowledge of an employer is therefore crucial in defining the extent of foreseeability of injury.

An employer may claim that the state of knowledge in relation to hand held vibrating tools means that it did not know, (nor ought to have known) that the vibration from tools could cause injury and nor could it have reasonably been expected to know the same. For example, until such injuries were generally reported within the industry an employer did not have the

2. [1968] 1 WLR 1776.

relevant knowledge of the risk.[3] Each case turns on its own facts but in *Hall* v *British Gas*[4] where the injury related to the vibration from using road breaking drills, it was held that the risk should have been known at the latest at 1975.[5] This was based on the accumulated scientific research at the time.[6] It was clearly known by this date that there was a risk of injury and so within a reasonable period of time thereafter precautions had to be taken. Further, in *Allen* v *BREL*[7] the employer should have undertaken surveys about HAVS by 1973. If this had been done then the claimant employee would have been offered medical advice by 1974 and he would then have been moved to work that did not expose him to the risk of injury by 1975. In the event the claimant left the defendant's employment in 1987 and suffered injury.

HSE and British Standard Guidance

Guidance from the Health and Safety Executive and the British Standards Institute has clarified what constitutes excessive exposure to vibration and helps identify the date when employers should have known of the risk of injury. The publications include:

(a) 1975: "DD43" the British Standard: (it developed a standard for exposure to injury caused by vibrating tools This included a categorisation of "acceptable/unacceptable" vibration).

(b) 1986: ISO 5349.

(c) 1987: BS 6842: Measurement of Evaluation of Human Exposure to Vibration Transmitted to the Hand.

(d) 1989: ISO 5349.

(e) 1991: BS 7482.

---

3. See *Hall* v *British Gas* [1998] Lawtel 7.4.98.
4. [*Supra*].
5. In *Armstrong* v *British Coal* [1996] Times 6 December unreported: the relevant date of knowledge of the risk of injury of the defendant was 1973 and precautions should have been taken by 1975.
6. "DD43" the British Standard: it developed a standard for exposure to injury caused by vibrating tools and was published in February 1975. This included a categorisation of "acceptable/unacceptable" vibration but was replaced by International Organisation for Standards (ISO 5349) in 1986 and BS 6842 in 1987. ISO 5349 1989 and 1991 BS 7482 and HSE "Hand Arm Vibration Syndrome 1994.
7. [2001] PIQR Q101.

(f) HSE "Hand Arm Vibration Syndrome 1994.[8]

Knowledge of the Employer: Case Law

In *Walker* v *Wabco Automotive UK Limited*,[9] the claimant suffered carpal tunnel syndrome from using vibrating tools as part of her assembly work for the defendant. At trial, liability was found on the basis of the Health and Safety Executive pamphlet HS(G)(60), published in 1990. On appeal it was held that although there was an obligation on the defendant to know of the guidance from the HSE, the guidance in the 1990 publication was too tentative to be relied upon. The pamphlet related to upper limb disorders generally and was not focused on the issue of HAVS. It did not establish that the defendant had fallen below the standard of the reasonable employer. A further contention was that the defendant could not have known of the risk of injury as there were no complaints of injury. In *Armstrong* v *British Coal Board* this was rejected and it was held:

> "It is trite law that an employer paying proper attention to the safety of his employees cannot rely on the absence of complaints from them as producing an impenetrable shield against possible liability".[10]

Other relevant case law on the issue of knowledge of the employer includes:

(a) *Joseph* v *Ministry of Defence* [1980] *Times* 4th of March 1980.

(b) *White* v *Holbrook Precision (Castings) Limited* [1985] IRLR 215.

(c) *McSherry* v *British Telcommunications plc* [1992] 3 Med LR 129.

(d) *Tremain* v *Pike* [1969] 3 All ER 1303.

These cases show that in hand-arm vibration syndrome cases defendant employers will find it difficult to argue that there was insufficient knowledge of the risk of injury after 1975. But in general where an employer keeps reasonably abreast of the developments in health and safety and acts upon them but could not have been expected to appreciate a particular new

8. These publications are closely analysed in *Armstrong* v *British Coal Corporation* [*supra*].
9. [1999] Lawtel 11.5.99 CA.
10. Per Judge LJ.

risk of injury, then there will be no liability. In assessing the issue of knowledge of the defendant, reliance will be placed on:

(a) Medical knowledge that had been disseminated to the industry at the relevant time.

(b) Health and Safety Executive and British Standards Institute publications (that are/were relevant at the time of the index allegations of negligence).

(c) Knowledge generated within the industry itself.

The categories of case are not closed and are determined on the particular facts: in *Heyes* v *Pilkington Glass Limited*[11] the claimant suffered a hand-arm syndrome allegedly caused by operating cranes in the course of his employment as a crane driver. He alleged that the vibration induced by the operation of the crane caused his condition. His appeal was dismissed. Whilst the industry was aware of the potential cause of hand-arm vibration syndrome from hand held tools, there was no such knowledge in relation to the operation of cranes. The defendant had adopted a recognised practice in operating the cranes which had been followed for a long time without injury. There had been no previous incidents in any of the other factories and so it was entitled to rely on its contention that it had no reason to know of the risk of HAVS in relation to the operation of cranes. This view was fortified by the fact that the HSE had not published any guidance on the dangers of operating electrically operated gantry cranes.

Breach of Duty

The Issues on Breach of Duty

In assessing the breach of duty, the issues in practice are:

(a) What measures ought the employer to have taken to minimise the risk of injury from exposure to vibration and when should these measures have been effected?

(b) Did the employer take reasonable care to safeguard the employee from the risk of injury due to vibration?

11. [1998] PIQR 303.

Particular Factors Relating to Breach of Duty

These issues can be further broken down in the following manner:

(a) *The Risk of Injury*

In *Armstrong* v *British Coal Corporation*[12] the claimant had to prove that the risk from the exposure to vibrating tools was in excess of that which a prudent employer would permit. But the standard required was "reasonable" care by the employer as there will always be some risks associated with most manual employment.

(b) *The nature and extent of use*

Breach of duty depended in part on the period over which the vibrating tools were being used, both in terms of daily use/frequency and cumulative period of use.[13] A causative connection had to be proved between the use of the vibrating tool and the injury. The more prolonged and frequent the use the more likely that causation is made out on the balance of probabilities.[14] In *Armstrong* v *British Coal Corporation,* the phrase used was "prolonged and regular exposure to vibrating tools".

(c) *Extent of Exposure*

Exposure did not have to be excessive but rather above a safe limit. In *Armstrong* v *British Coal Corporation,* it was held that the British Standard (DD43) was too low and so the defendant could not "shelter behind it". Reasonable exposure levels were those contained in the 1994 HSE booklet on Hand-Arm Vibration Syndrome. Exposure in excess of that prescribed exposure meant the job had to be rotated. The risk of injury was not so high as to warrant a prohibition on the use of such vibrating tools. The appropriate level of exposure was that commensurate with the industry and any exposure in excess of that advised by the handbook would require job rotation.

---

12. [1996] *Times* 6 December.
13. Defendants will want to make an assessment of the actual "anger time" which is the precise time the tool is used. This will normally be shorter than the period of the employee's shift.
14. The issue was fully covered in *Hall* v *British Gas Plc* [*Supra*]: note that this technical information on the extent of the vibration ought to be obtained from the manufacturer.

(d) *Action in respect of the tools*

Based on the date of knowledge of the risk of injury, the employer has to take reasonable steps to reduce that risk. If the employer recognised (or ought to have recognised) the risk of injury but failed to do anything about it then a breach of duty would probably result. Moreover, practical precautions have to be actively considered on a continuing basis.[15]

(e) *Investigations*

The employer has to make relevant investigations of the potential risk of injury from vibrating tools. In larger companies, with a big workforce at risk, the duty imposed is higher and requires active research on the point.

(f) *Advice, warnings and training*

The employer must give warnings to an employee on the dangers of using vibrating tools. Any such warning has to be adequate and considered within the practical reality of the industry in which the individual was working. For example, with HAVS, where the blanching is at stage 1 it can still be reversible and so an appropriate warning was:

> "If you are working with vibrating tools and you notice that you are getting some whitening or discolouration of any of your fingers then in your own interests you should report this as quickly as possible. If you do nothing you could end up with some nasty problems in both hands."[16]

Each case turns on its facts but the necessary action required as to training and advice will include:

- relevant literature produce by medical officers or safety officers;
- talks, seminars and courses on which the employees are sent on the issue of hand-arm vibration syndrome;[17]

---

15. Evidential issues include: (a) whether the vibration levels are checked (b) whether a company policy has been created (c) whether the vibrations were reduced at source. See *Hall* v *British Gas Plc* [*supra*].
16. See *Armstrong* v *British Coal Corporation* [*supra*].
17. Note that wearing gloves does not on the medical evidence have any effect on the risk of injury. *Transco Plc* v *Griggs* [2003] EWCA Civ 564.

- Job rotation and limiting exposure to vibrating tools is an important consideration;[18]

- Instruction and supervision at the workplace is a material requirement including the level of training of the management. How a defendant puts these requirements into practice is a matter for it alone as long as it complies with its duty of care.[19]

(g) *Precautions*

A system of regular medical examinations must be considered. In *Hall* v *British Gas Plc*,[20] the employer asserted it was difficult to administer such a scheme and in any event employees were reluctant to own up to an injury that could reduce their earnings. This was rejected by the judge as the previous studies had shown that immediate action is required to deal effectively with this condition. A defendant that failed to comply with this duty would invariably be in breach of duty as it was contrary to the actions of a reasonable and prudent employer having due regard to the welfare of the employees.

> "Had the defendants whether in 1977 or 1980 issued warnings then no doubt the claimant in the early 1980s would have realised what was happening and then no doubt have reorganised work patterns so that Mr. Smith could work without enduring further symptoms. Alternatively if it were not possible they would no doubt have dismissed him. As it was the claimant continued until and after his symptoms had reached a stage where the damage was irreversible."

(h) *Statutory duty*

The main statutory duties imposed will be the Provision and Use of Work Equipment Regulations 1992 and 1998 in addition to the Workplace (Health,Safety and Welfare Regulations) 1992.

18. *Armstrong* v *British Coal Corporation* [*supra*].
19. *Armstrong* v *British Coal* [*supra*]: Per Judge Stephenson.
20. [*Supra*].

Expert Evidence

It is essential to obtain proper expert evidence on the issue of breach of duty and causation. Relevant expertise is necessary and so a consultant engineer's report will be required.

Expert Engineering Evidence

The report will have to deal with the following issues:[21]

(a) The tasks undertaken by the claimant, based on his statement.

(b) The period and extent of exposure, both in terms of daily use and the lifetime during the employment.[22]

(c) Identification of the relevant guidelines (HSE, BS and ISO).

(d) The nature and extent of the vibration caused by the tools and whether this breached the accepted guidelines. Reference to the manufacturer must be considered and other available data relating to a similar process. But also an assessment is needed by reference the guidelines (HSS, BS ISO) and an opinion on whether they are breached.

(e) An opinion is required as to whether the exposure was sufficient to cause HAVS. Daily dose recommendations based on HSE publications and other guidelines need to be considered in addition to frequency and magnitude of exposure.

(f) Whether and in what manner the system of work was negligent including, the date of knowledge of the risk of injury of the defendant.

(g) Causation: and specifically whether there are any other factors in the system of work or the work history that could have caused the condition.

Expert Medical Evidence

In addition medical evidence is required primarily from a vascular surgeon, but it depends on the nature of the injury. The

21. This is not an exhaustive list.
22. See *King* v *Samsung Heavy Industries* [2002] Lawtel 10.4.02: *Transco Plc* v *Griggs* [*supra*].

injury may require an orthopaedic or rheumatological opinion (e.g. Carpal tunnel syndrome) and relevant issues include:

(a) When did the symptoms present and does this tie in with the use of the tools at work?

(b) Whether the extent of exposure can account for the severity of the symptoms on a medical basis.

(c) Assuming that the above are satisfied then if the claimant had been removed would there have been a full recovery or would some conditions have presented in any event? If so what was the cut off date for irreparable harm being done (if relevant).

(d) Causation: is the index injury caused by the allegations of breach of duty that the claimant makes? Are there other potential causes of the condition?

(e) Condition and prognosis, including an assessment of the progression and the claimant's ability to work in the future on account of the index injury.

Causation

General Guidance

The claimant has to prove the causal connection between the alleged breach of duty and the injury suffered. The burden of proof never shifts on to the defendant.[23] Accordingly the considerations are:

(a) Has the employer's breach of duty caused or materially contributed to the claimant's injury?

(b) If so, then to what extent did it cause or contribute and specifically:

- Should the employee's injury be apportioned between the injury attributable to breach of duty and injury attributable to other causes and if so, how?

23. *Holtby* v *Brigham and Cowan Limited* [2000] *supra.*

- Has the breach of duty accelerated the index symptoms and if so, for what period?
- What was the last date that a cessation of use of the vibrating tools would have led to a full or substantial recovery?
- What damage would have been suffered in any event?[24]

Chapter 2 deals in detail with the principles of causation in this type of injury. As to case law on hand-arm vibration syndrome, it was noted in *Armstrong* v *British Coal Corporation* that the claimant had first to prove a material contribution to the injury caused by the breach of duty:

> "it must be a matter of common sense given that exposure is cumulative that the probability is that failure to reduce exposure by rotation materially contributed to the symptoms."

Case law: Allen v British Rail Engineering Limited[25]

An appeal on quantum related to HAVS suffered by employees who used vibrating tools. The claimant had worked for the defendant since the late 1950s and suffered HAVS since 1968. It was held that surveys to assess HAVS should have been undertaken by the defendant by 1973. The questions posed for the Court of Appeal were:

(1) Was the judge right to rule that the employer was liable only for the excess damage attributable to the negligence, or was the employer liable for all the damage?

(2) Was the judge entitled to conclude that the claimant would have suffered some damage even if there was no negligence by the employer and this damage amounted to half of the total damage suffered?

The claimant argued that once a vibrating tool was proved to make a material contribution to the injury then full compensation was payable for the entire injury. In contrast, the defendant contended it was liable only for its actual contribution

24. Causation arguments based on the "vulnerability to injury" caused by vibrating tools probably do not apply: see *Hall* v *British Gas* [*supra*]. *Transco Plc* v *Griggs* [2003] EWCA Civ 564.
25. [2001] PIQR 101.

to the injury suffered by the employee. In principle, it asserted this was consistent with the exclusion of liability for further injury suffered both pre and post employment with the defendant.[26] The Court of Appeal upheld the trial judge's apportionment of 50% which was the best she could have done in the circumstances. It was based on the life time exposure which was then attributed in percentage terms, to the percentage proportion of damage done before liability was established and after he left the defendant's employment (assessed at 40%). General damages were reduced to reflect this apportionment.

Further the Court of Appeal set down the following five propositions applicable to this type of case:

(a) To succeed on liability the claimant has to prove that the employer's tortious act caused a material contribution to the employee's disability.

(b) There can be cases where the injury was caused wholly by the joint negligence of two tortfeasors (a head on car collision where the claimant is the passenger in one car, where both cars are driven negligently).

(c) But in principle the amount of the liability will be limited to the extent of the contribution which the employer's tortious conduct made to the employee's disability.

(d) The court will apportion as best it can but should not deny the claimant relief on the basis that he cannot demonstrate accurately the proportion of injury caused by one employer.

(e) The evidence to be called should be sufficient to enable a court to make a just apportionment and should be proportionate to the amounts at stake.

Case Law: Hall *v* British Gas[27]

In this case, the defendant argued that the employee was known to be susceptible as he developed the symptoms of HAVS and so the damages ought to reflect this fact. This assertion was rejected. The fact that an employee developed hand-arm vibration syndrome after years of use of a vibrating

26. Reliance was placed on *Bonnington Castings Limited* v *Wardlaw* [1956] AC 613, *Thompson* v *Smiths Ship Repairers (North Shields) Limited* [1984] QB 405 and *Holtby* v *Brigham and Cowan (Hull) Limited* (2000) 3 All ER 421: all discussed in Chapter 2.
27. [*Supra*].

tool did not make him any more or less susceptible to injury than any other employee. It depended on the nature and extent of the use of the tool and as these factors varied then no valid comparison between employees and their individual susceptibility to injury can be made.

In addition, the judge rejected the defendant's contention that the alleged susceptibility to injury meant that the injuries should be treated as acceleration injuries. The defendant's comparison to latent pre-existing conditions such as osteoarthritis in a whiplash victim was not well drawn. There was no evidence in this case of a pre-existing injury and no evidence of when that pre-existing condition would have arisen in any event. The defendant's comparison on causation to *Thompson* v *Smiths Ship Repairers (North Shields) Limited*[28] was inappropriate as it related to industrial deafness. There was medical evidence in that case that the hearing of the employee was already damaged prior to the breach of duty, which was not the case here. It was held there was a material contribution to the claimant's condition and there was no evidence that could be relied upon to reduce the damages.

Finally, there was no apportionment of the damages for the injury caused by the culpable exposure and the non-culpable exposure, as there was no evidence that the claimants had suffered latent damage prior to the period when the breach of duty occurred (i.e. at the date of the defendant's knowledge). If there was such evidence then there would have to be an apportionment (as in *Armstrong*), but in this case that was not so.

Case Law: Smith v Wright Beyer[29]

The issue here was non-negligent exposure and causation: the defendant claimed a reduction on the damages to take account of alleged non-negligent exposure of the claimant for which it had no liability. The defendant contended that the pre-1977 exposure must have had some effect that would have presented in symptoms to an extent in any event. Alternatively, that in reality the claimant would have taken the risk of injury anyway. This was rejected by the Court of Appeal, because in 1977 if the defendant employer had discharged its duties as employer then notwithstanding the long history of use up to the point of his employment, the claimant would not have developed the

28. [1984] QB 405.
29. [2001] EWCA Civ 1069.

condition. No further significant exposure would have taken place. The defendant employer could have called evidence that practically such precautions and reorganisation of work practices could not have been implemented but there was no such evidence. No apportionment arose.

Case Law

Transco Plc v *Griggs*.[30] The claimant succeeded at first instance in proving he suffered PAD in the course of his employment. It was held the PAD (Palm Arch Disease) was caused by his use of vibrating tools. On appeal by the defendant it was held that it was important for the trial judge to assess the expert evidence and decide which was more reliable. As the expert evidence was to the effect that it was 85% likely that the vibratory tools caused "this sort of injury" then the claimant ought to succeed. Once the degree of exposure, breach of duty and causation was established then it would be wrong not to find for the claimant.

Quantum

The JSB Guidelines on HAVS have been held to be too low as they do not acknowledge the severity of the condition and the disability which it caused.[31] Notwithstanding the same, there is set out below a series of cases that provides an indication of general damages. In addition a claimant is entitled to claim for any future loss of earnings or loss of pension if he cannot work. The claimant may be able to return to alternative but equally well paid work but nevertheless still be handicapped in the labour market.

Disadvantage on the labour market

In *Armstrong* v *British Coal Corporation*,[32] it was noted that a hand-arm vibration injury to a miner placed him at a disadvantage in the labour market and this point has wider application for all claimants whose work is manually based. But it was also highlighted that other serious disabilities had to be taken into account if relevant to the ability to undertake manual work. If that is so then HAVS at the less serious level will not have such a profound impact on a claimant's employment prospects. The awards made under this head are contained in the table set out below.

30. [2003] EWCA 564.
31. *Armstrong* v *British Coal Corporation* [*supra*].
32. [*Supra*].

Quantum

Case Name	Nature of Injury	Award
Hall v *British Gas Plc* [1998] Lawtel 7.4.98	Male (42), Exposed to excessive vibrating tools: suffered HAVS. Tingling and numbness in the hands up to the forearms. Whitening of fingers. Stage 3. Could not undertake activities in cold and wet weather. Condition permanent. Could still work as a technician but handicapped in labour market.	Award: General Damages: £15,000. Handicap in the LM: £15,000
Davies v *British Gas Plc* [1998] Lawtel 7.4.98	Male (36) handled pneumatic drills and sustained HAVS. Whitening of fingers, numbness and tingling in all fingers and thumbs: Stage 3(R)(5)/3L(5): severe frequent attacks. Could no longer labour outside. Permanent and was handicapped in labour market by injury and partial loss of sight.	General Damages: £15,000. Handicap in LM: £25,000
Leyland v *British Gas Plc* [1998] Lawtel 7.4.98	Male (36) sustained HAVS. Attacks of whiteness spread into fingers and hands. Stage 3(4): severe frequent attacks. Medically retired in 1996, but other factors relevant. Could not drive as no grip and he required assistance to dress. Permanent condition.	General Damages: £20,000. Handicap in LM: £2,500

Case Name	Nature of Injury	Award
Earle v *British Gas Plc* [1998] Lawtel 7.4.98	Male 52, suffered HAVS symptoms in 1984 with numbness and whitening of fingers. Stage 3/severe. Retired on account of hypertension. Lost ability to grip or lift. Permanent, but also a loss of life expectancy (which was unrelated to HAVS).	General Damages: £15,000
Bonney v *British Gas PLC* [1998] Lawtel 7.4.98	Male (34) suffered HAVS in using vibrating tools. Symptoms started in 1992. Stage 2: occasional attacks of moderate nature affecting one or more fingers. Loss of sensation in cold weather. Continued to use tools and condition could deteriorate.	General Damages: £9,000. Handicap in LM: £20,000
Allen v *BREL* [2002] PIQR Q101	Suffered HAVS whilst using vibrating tools from the late 1950s to 1987. Permanent condition but some of the condition was attributable to injury caused prior to the defendant's breach of duty.	General Damages: £9,000. this was reduced to £4,000 to take account of the pre-existing injury, or that injury he would have suffered in any event.

Case Name	Nature of Injury	Award
Griggs v *Transco Plc* [2002] Lawtel 5.7.02	Male (53): employed for 24 years as a road worker and used pneumatic tools known as "road breakers". Suffered HAVS, which affected his blood flow, blanching of the fingers and could not continue his employment.	General Damages: £16,000
Smith v *Wright and Beyer Limited* [2001] EWCA Civ 1069	Male (53) employed as a boilermaker since 1961 and he used vibrating grinding tools. Developed HAVS in "severe not modest category": ability to work and enjoy daily activities was substantially compromised.	General Damages: £14,000. Handicap in the LM: £3,000

Summary of the Chapter

Set out below is a summary of this chapter in addition to a series of points contained within the schedule that covers the issues that practitioners have to consider when bringing or defending these claims:

Issue	Claimant's Perspective	Defendant's Perspective
Duty of Care: At common law based on *Stokes* v *Guest et al.* Statutory regulation in part from Provision and Use of Work Equipment Regulations 1992/98 and the Workplace (Health, Safety and Welfare) Regulations 1992.	Obtain a statement from the claimant at an early stage stating the precise mechanism of the injury: magnitude of exposure, frequency and dose levels.	Obtain the claimant's version of events and basis of case as soon as possible. Tie the claimant down to a version of events as early as possible.
Foreseeability: Knowledge of the employer of the risk of injury: case law covers the point but the cut off date after which foreseeability cannot realistically be denied is about 1975.	Does the defendant concede the foreseeability point based on knowledge of risk of injury?	Consider whether this is a relevant point in issue or whether it can be conceded. This depends on the type of machinery and the nature of system of work but invariably it is now difficult to contest.
Breach of Duty The breach of duty at common law will depend on the risk of injury, the magnitude of that risk, the steps taken to prevent over-exposure, the warnings given and the precautions taken.	The claimant has to provide detailed instructions and this statement will form the basis of the expert engineer's report. Expert evidence on this point is essential. Did the system of work cause the injury?	Expert evidence has to be obtained on the system of work. Full statements from management setting out the precautions taken. Statements should be obtained from colleagues as to whether they had problems. Were complaints made and if so why and when?

Issue	Claimant's Perspective	Defendant's Perspective
Medical evidence: Causation Probably vascular surgery rheumatological as opposed to orthopaedic.	Medical evidence must cover causation: consider pre-existing symptoms. Also the last date that removal of tools meant a full recovery. Assess the likely level of disability in any event from safe use. Assess the connection between the appearance of symptoms and the work. Clarify the future ability to work and/or extent of handicap in the labour market. Consider *BREL* v *Allen* on causation.	Expert medical opinion needs to address causation: whether there is an acceleration or exacerbation injury. Consider DSS Benefits Agency notes and/or medical notes to assess first presentation of the symptoms. What is the relevance of non-negligent exposure (see *Smith* v *Beyer*). Was there a series of causes such that an attribution can be made?
Limitation:	Is limitation a material consideration?	Assess all medical and DSS Benefits Agency records to assess the date of knowledge of injury of the claimant.

Section 2: The Medical Perspective

Overview

Adverse effects caused by vibration in the hands and arms have been recognised since 1911, when Loriga reported dead fingers in Italian miners using pneumatic tools such as percussion hammers and drills.

The characteristic health effect is known as hand-arm vibration syndrome ("HAVS"), a syndrome of impaired circulation and nerve function in the hands and arms. It was formerly known as "vibration white finger" ("VWF"), "vibration-induced white finger" ("VIWF"), "trauma-induced vaso-spastic disease" and "occupationally related Raynaud's phenomenon".

The types of tools most commonly associated with HAVS are handheld vibrating pneumatic tools, for example drills, sanders, rivet guns, jack hammers, grinders, screwdrivers and impact wrenches. Users of chainsaws, hedge cutters and grass trimmers are also at increased risk, and forestry workers may now be the leading source of new cases.

Basic Anatomy

The blood supply to the digits is provided by small arteries which branch to form "arterioles",[33] and then capillaries, the smallest vessels, which deliver oxygen and nutrients to the tissues (see figure 7).

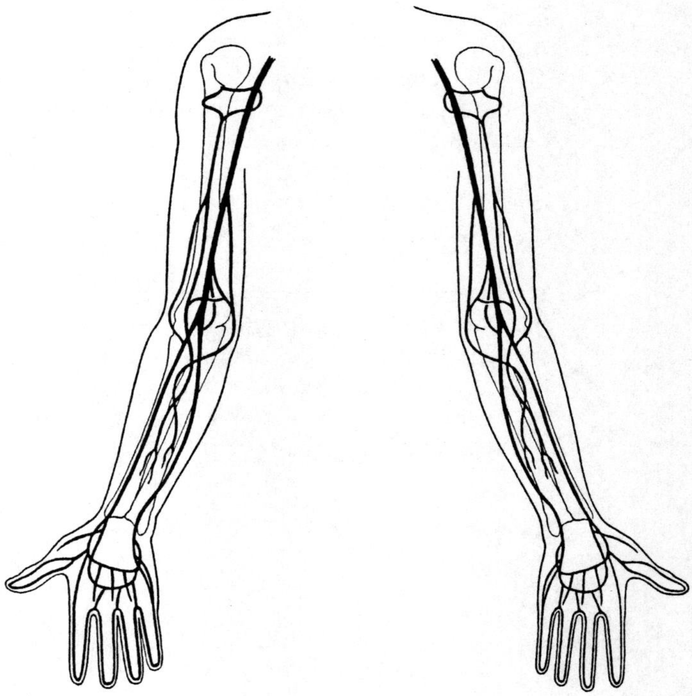

Figure 7
The arterial supply of the upper limbs

The walls of the arteries and arterioles contain smooth muscle cells, which are capable of contracting or relaxing in an involuntary manner, in response to a number of stimuli. This affects the "lumen" (calibre) of the vessels, and thus the blood flow within them (see figure 8).

33. The smallest arteries.

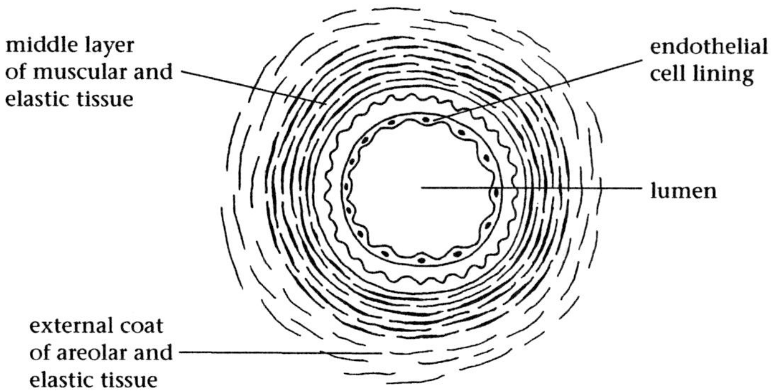

Figure 8
Cross-section of a small artery

The normal response of these blood vessels to cold, namely constriction, is partly mediated through a branch of the involuntary nervous system known as the "sympathetic system".[34]

The walls of small arteries and arterioles are lined by cells known as "endothelial" cells, which also play an important role in the regulation of blood flow. Injury of the endothelial cells, for example through cold, results in local aggregation of blood constituents known as platelets. The aggregated platelets release a chemical mediator,[35] which stimulates the adjacent smooth muscle cells to contract.

This chemical contractile system is checked and balanced by other chemical mediators produced in the endothelial cells,[36] which inhibit the contraction of the smooth muscle cells. In practice this means that once the stimulus for contraction has passed, the blood vessels should revert quickly to their normal state.

34. The activation of the sympathetic nerves supplying these blood vessels, results in the release of epinephrine and related chemical mediators, which stimulate receptors on the surface of the smooth muscle cells. This in turn triggers the release of calcium stores within the muscle cells, leading to their contraction.
35. 5-hydroxytryptamine, "5-HT".
36. Including prostacyclin and endothelium-derived relaxing factor.

Basic Physics

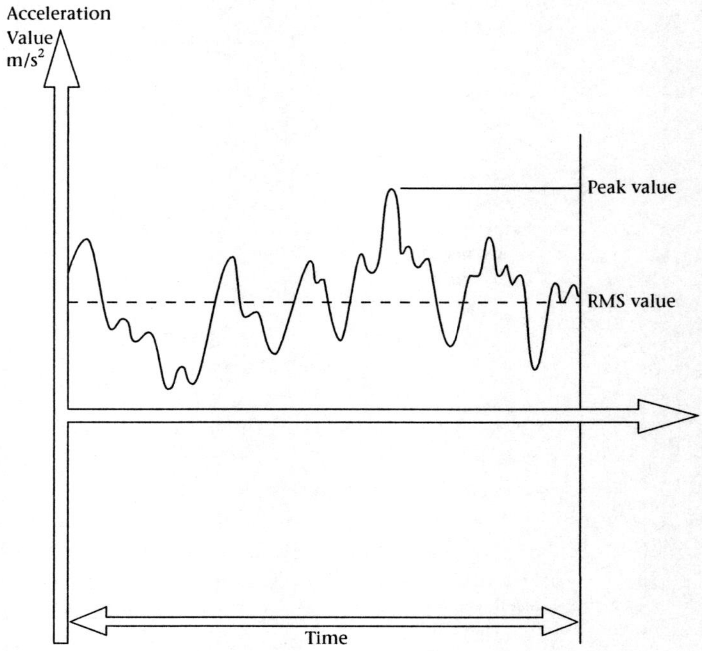

Figure 9

Vibration is an oscillatory motion arising from a source, which is usually mechanical in origin. The important components of the vibratory wave are (see figure 9):

(a) The *amplitude*: the distance in metres through which the vibrating surface moves each time it vibrates.

(b) The *velocity*: the speed at which the surface is moving at any instant as it vibrates, expressed as the number of metres travelled in each second (m/s).

(c) The *acceleration*: the rate at which the speed of the vibrating surface is increasing at any instant, expressed as the increase in speed (m/s), in each second (m/s^2).

(d) The *peak value*: the maximum acceleration measured during a period of time.

(e) The *root mean square* ("RMS") value, a method for providing an average figure for acceleration over a measured period of time.[37]

(f) The *frequency* of the wave: the number of times a complete cycle of motion is repeated each second. One cycle per second is one "hertz" (Hz).

(g) The *vector*: whether the displacement of the vibrating surface is up and down, back and forth or side to side.[38]

(h) The *degree* of any additional rotational movement of the surface as it moves up and down, back and forth and from side to side.[39]

Analysis of vibration is not easy. There may be multiple vibrations in existence at any one time, some of low frequency, and others of high frequency. The vibrating surfaces may be accelerating at different rates according to whether the movement is measured up and down or from side to side.

These problems are overcome by using an instrument known as an accelerometer (a vibration transducer), which measures the acceleration of the vibrating surface in each direction (up and down, back and forth and side to side), across a range of frequencies.

The average (RMS) acceleration values are then given a "weighting" according to how important the frequency of vibration is. This is because at certain critical vibration frequencies, there is greater transfer of vibration energy to human tissues.

This is the phenomenon of "resonance", where the tissues begin to vibrate at the same frequency as the source. For the hand, the principal resonance frequencies are about 30–150 Hz, and for

37. The formula is complex. The figure for acceleration at each instant is measured, and multiplied by the same figure ("squared"). All of the squared figures for acceleration over a measured time period are added together, and an average obtained. The square root of this figure is the RMS value.

38. By convention these different types of motion are referred to as being in the direction of the "X","Y", "Z" axes.

39. This is referred to as "pitch", "roll" and "yaw", just as in the motions of a boat or aeroplane.

the lower arm about 15–30 Hz. In practice this means that frequencies of 150 Hz are more likely to injure the hand and less likely to injure the lower arm.

When the frequency weighted acceleration of a particular machine is established, standard tables can be used to determine the maximum acceptable daily dose of vibration. Symptoms of HAVS do not usually occur with normal tool usage if the frequency weighted acceleration is below about $1 m/s^2$ RMS.

However, more recent studies have suggested that impact vibration (high impulsiveness) may be an independent risk factor distinct from continuous vibration. Examples of impulsive tools include jack hammers, electric hammers and impact guns.

Basic Pathophysiology

It is generally thought that the mechanical stimulus of vibration causes abnormal growth (hypertrophy) of the muscular wall of small arteries and arterioles. This ultimately results in narrowing of the lumen, reduced blood flow, and damage to the endothelial cells.

Associated factors include:

(a) An exaggerated response to cold, mediated by abnormal activity of the sympathetic nervous system. This causes the typical blanching induced by cold.

(b) Leakage of fluid into tissues, resulting in swelling of the affected digits.

(c) Intermittent interruption of the blood supply (ischaemia), which damages the myelin sheaths[40] of nerves supplying sensation and power to the affected hand and digits.

(d) In severe cases, an accumulation of scar tissue (collagen).

40. The white tissue covering a nerve fibre, which acts in a similar way to the insulation covering a wire.

Clinical Features

Symptoms

There is usually a "latent interval" between first exposure to vibration and symptoms. This period may vary considerably due to individual factors, such as cigarette smoking.

The symptoms are initially precipitated in cold, wet conditions, but with continued exposure to vibration the temperature change required to induce symptoms falls, and may even occur in the summer, in wet and windy conditions. The characteristic symptoms of circulatory impairment form the so-called "Raynaud's phenomenon":

(a) Initial painless blanching of digits ("dead finger" or "white finger");[41]

(b) A blue "cyanotic" phase;

(c) A red, painful phase ("hot aches"), due to abnormally increased blood flow (hyperaemia).

Other symptoms may include:

(a) Swelling of the affected digits;

(b) Tingling and pins and needles due to associated nerve damage;

(c) In more severe cases, impaired sensation to a number of modalities, including temperature, vibration and touch;

(d) Impaired grip strength and loss of manual dexterity.

Classification of HAVS

The *Stockholm Workshop Scale* is used widely to classify the symptoms of HAVS. Each hand is staged separately, in terms of the overall effects on circulation (Raynaud's phenomenon), and the overall effects on nerve function (sensori-neural effects).

41. The blanching is initially confined to the tips of one or more fingers, but progresses with continued exposure to vibration. The thumbs are usually the last digits to be affected.

Stage	Grade	Description
Cold-induced Raynaud's Phenomenon		
0	–	No attacks
1	Mild	Occasional attacks affecting only the tips of one or more fingers
2	Moderate	Occasional attacks affecting distal and middle (rarely also proximal) phalanges of one or more fingers
3	Severe	Frequent attacks affecting all phalanges of most fingers
4	Very severe	As in stage 3, with trophic skin changes (gangrene) in the finger tips
Sensori-neural effects		
0_{SN}	–	Exposed to vibration but no symptoms
1_{SN}	–	Intermittent numbness, with or without tingling
2_{SN}	–	Intermittent or persistent numbness, reduced sensory perception
3_{SN}	–	Intermittent or persistent numbness, reduced tactile discrimination and/or manipulative dexterity

Each hand is staged separately for Raynaud's phenomenon and sensorineural effects, and the number of affected digits is indicated in each hand. For example, severe Raynaud's phenomenon affecting both hands, with three affected digits on the left, and two on the right will be: 3L(3)/3R(2).

Diagnosis

The diagnosis is usually made on the basis of the characteristic symptoms, exposure to HAV, and the exclusion of other causes of Raynaud's phenomenon.

Other causes of Raynaud's phenomenon include:

(a) Primary Raynaud's disease (constitutional white finger);

(b) Degenerative disease of peripheral blood vessels, in conditions such as diabetes mellitus;

(c) Connective tissue diseases, such as scleroderma;

(d) Drugs used to treat high blood pressure such as beta-blockers;

(e) Sometimes carpal tunnel syndrome, which is also associated with HAV, may cause initial diagnostic confusion.

There is no single diagnostic clinical test. The tests that are available have not been standardised. The tests which may be utilised include:

(a) Measurement of blood flow and pressure in the limbs and digits;[42]

(b) Measurement of finger systolic blood pressure before and after cold air or water provocation;

(c) Measurement of digital blood flow[43] before and after cold air or water provocation;

(d) Measurement of skin temperature before and after cold air or water provocation;[44]

(e) Tests of nerve function, such as two-point discrimination, finger tip vibration threshold, hot-cold perception, grip strength and nerve conduction studies.

HAVS is a prescribed industrial disease (PDA11) when the necessary diagnostic and work criteria are met.

Prevention

Pre-employment medical screening should identify workers with a history of symptoms of HAVS who may be at increased risk. All workers exposed to HAV need a programme of active

42. Using Doppler ultrasound scanning.
43. By a technique known as "venous occlusion plethysmography".
44. The cold provocation tests are used to verify Raynaud's phenomenon on cold exposure, delayed return of blood flow, and reactive hyperaemia.

education and training with regard to the risks and methods of prevention.

They need to be aware of the need to service tools regularly; to wear protective clothing provided; and to report symptoms as soon as they occur in between scheduled examinations.

As HAVS is related to the accumulated dose of vibration received over time, control measures should be directed to reducing the source of the vibration, and reducing transmission by isolating and damping techniques.

Many items of equipment, such as chainsaws, have now been designed with anti-vibration features, including higher power to weight ratios, and non-slip surfaces, which reduce the need for a tight grip.

Anti-vibration gloves, which can be worn as inserts inside conventional gloves, are likely to be of limited benefit. Mitts should be worn in preference to gloves where possible, as they offer more effective protection against cold.

Treatment

The mainstay of treatment is avoidance of further HAV, avoidance of cold exposure, and maintenance of the core body temperature. Cessation of smoking is always advised, since chemicals in cigarette smoke have adverse effects on the blood flow in peripheral vessels.

Drug therapy is sometimes indicated.[45] However, since these drugs affect most blood vessels in the body, and not just those in the hands and limbs, some patients find the systemic side effects intolerable. These include dizziness, flushing, headache and impotence. The drugs may need to be taken for several years.

If avoidance of HAV is impossible, a modified work routine may be necessary, with reduction of HAV exposure time, and instructions to grip tools as lightly as possible. HAVS is probably reversible, except in the most severe cases. However, the sensori-neural component usually takes longer to recover than the vascular phenomena.

45. With agents such as calcium channel antagonists, which act as vasodilators.

Further Reading

Pelmear, P. Vibration "Hand-arm and Whole Body" in *Hunter's Diseases of Occupations (9th edition)*, London, Hodder, 2000.

Jones, C. "Occupational Hearing Loss and Vibration Induced Disorders". In: *ABC of Work Related Disorders,* London, BMJ Publishing Group, 1997.

CHAPTER 10
OCCUPATIONAL STRESS RELATED ILLNESS

SECTION 1: THE LEGAL PERSPECTIVE

Overview

The main legal issues as to work related stress injuries are:

(a) whether there has been a breach of duty by the employer;

(b) if so, whether some injury was foreseeable; and

(c) whether and to what extent the negligent system of work has caused the employee's injury.

Proof of Injury

To succeed in a work related stress injury claim the initial point of entry is to assess whether a recognised "injury" has been suffered that is, one for which the law provides a remedy in damages.[1] If there is no proven injury the claim will fail and accordingly the medical evidence addressing injury and causation must be the starting point for bringing or defending a claim.

In general, in order to appreciate the nature of stress related injuries in the workplace the following publications provide assistance. They also provide an indication of the knowledge of an employer of the risk of injury:

(1) Law Commission Consultation Paper: *Liability for Psychiatric Illness*: [LCCP No.137 1995]

1. "Injury" is defined within s.38 of the Limitation Act 1980: as including *"any disease and any impairment of a person's physical or mental condition"*. There is no need to prove that a claimant was of reasonable robustness: *Page* v *Smith* [1995] 2 All ER 736.

(2) Law Commission Consultation Paper: *Liability for Psychiatric Illness*: [LCCP No.249 1998]

(3) Health Education Authority: *Stress in the Public Sector-Nurses Police, Social Workers and Teachers* [1988].

(4) Education Service Advisory Committee of the Health and Safety Commission: *Managing occupational stress: a guide for managers and teachers in the schools sector* [1990]

(5) Health and Safety Executive: *Stress at Work* [1995]

The Legal Principles

Case Law Prior to Sutherland v Hatton

Prior to the case of *Sutherland* v *Hatton,*[2] the law relating to the duty on an employer to prevent this type of injury was set out in *Walker* v *Northumberland County Council.*[3] A social worker, was responsible for four teams of social services fieldworkers, who were employed by the defendant council. He suffered a mental breakdown in 1986, prior to which he had been subjected to significant overwork. When he returned the system of work was not changed and he suffered a second breakdown. Eventually, he was dismissed on the grounds of permanent ill-health in 1988.[4]

The court held first, that a psychiatric injury was within the scope of an employer's duty to provide a safe system of work. Secondly, that the proven duty of care meant that an employer has to act to protect the employee in relation to a risk of harm, but not risks that are negligible. The practicability of the remedial measures is a material factor. Whilst initially it was not reasonably foreseeable that injury would be caused by exposing the claimant to an excessive workload, the same could not be said in relation to the second breakdown. The claimant had been exposed to a risk that was greater than the risk which would have impacted upon a manager in that type of job. Further, the

2. [2002] IRLR 263; [2002] 2 All ER 1.
3. [1995] 1 All ER 737.
4. As to the general principles relating to "nervous shock" cases which have relevance to the imposition of a duty of care: see *Page* v *Smith* [1996] 1 AC 155 and [1995] 2 All ER 736. *Frost* v *Chief Constable of S. Yorkshire* [1998] 3 WLR 1509, *Alcock* v *Chief Constable of South of Yorkshire* [1992] 1 AC 310.

employer should have taken measures to reduce the workload by additional assistance or restructuring.

Sutherland v Hatton

Sutherland v *Hatton* is now the benchmark case in this area of the law. All previous case law has to be considered in the context of *Sutherland* v *Hatton*. In determining the standard of the duty of care the Court of Appeal in *Sutherland* made two broad points:

(1) "The law of tort has an important function in setting standards for employers as well as drivers, manufacturers, healthcare professionals and many others whose carelessness may cause harm. But if the standard of care expected is set too high or the threshold of liability too low there may be unforeseen and unwelcome effects upon the employment market"

(2) "Some things are no-one's fault. No one can blame an employee who tries to soldier on despite his own desperate fears that he cannot cope, perhaps especially where those fears are groundless. No-one can blame an employee for being reluctant to give clear warnings to his employer of the stress he is feeling...The key is to offer help on a completely confidential basis. The employee can then be encouraged to recognise the signs and seek help without fearing its effects on his job or prospects".[5]

It was also held that there must be a recognised psychiatric illness. Once such an injury is proved then it will be treated in law as no different from any other physical injury or illness, but with added control mechanisms. The Court of Appeal expressly approved the following passage from the Law Commission consultation paper on "Liability for Psychiatric Illness:"[6]

> "On the other hand medical and legal experts working in the field, who are the people who most commonly encounter those complaining of psychiatric illness have impressed upon us how life shattering psychiatric illness

5. *Sutherland* v *Hatton* [*supra*].
6. LCCP No.137 1995 at 1.9.

can be and how in many instances it can be more debilitating than physical injuries."

The Duty of Care and Foreseeability

The Nature of the Duty of Care

Employers owe their employees a duty to ensure they are reasonably safe in the course of their employment.[7] In psychiatric injury cases that duty may be subdivided between "primary" and "secondary" victims. The distinction was made in the following manner in *Sutherland*:

> "A primary victim is usually someone within the zone of foreseeable physical harm should the defendant fail to take reasonable care...A secondary victim is usually someone outside that zone; typically such a victim foreseeably suffers psychiatric harm through seeing, hearing or learning of physical harm tortiously inflicted on other."[8]

The issues relating to secondary victims will not apply in work related psychiatric injuries. But the ordinary principles of employer's liability do apply in a case of psychiatric injury caused in the course of a claimant's employment.[9] Against that background *Sutherland* v *Hatton* set out the categories of claim for psychiatric injury:

(a) Tortious claims by primary victims, are usually within the foreseeable scope of physical injury, but also in some cases of psychiatric harm.

(b) Tortious claims by secondary victims, for example witnesses to the harm caused to others.

(c) Contractual claims by primary victims, where the harm suffered is reasonably foreseeable as arising as a result of the employment relationship.[10]

7. *Wilsons and Clyde Coal and Co Limited* v *English* [1938] AC 57.
8. At para 19: see also *Page* v *Smith* [*supra*]. *Alcock* v *Chief Constable of South Yorkshire Police* [1992] 1 AC 310. *Frost* v *Chief Constable of South Yorkshire Police* [1999] IRLR 110. [1998] 3 WLR 1509.
9. *Walker* v *Northumberland County Council* [1995] IRLR 35. [1995] 1 All ER 737.
10. See *Garrett* v *London Borough of Camden* [2001] EWCA Civ 395, *Petch* v *Commissioners of Customs and Excise* [1993] ICR 789. *Walker* v *Northumberland CC* [*supra*]. *Foumeny* v *University of Leeds* [2003] EWCA Civ 557.

(d) Contractual claims by secondary victims. The harm is suffered to a third party which in turn causes foreseeable injury to the claimant within the context of a contractual relationship with the defendant (e.g. policemen).

The Threshold of Foreseeability

A central issue is whether an employer should have foreseen the risk of injury to a particular claimant employee as caused by the system of work. Further, the steps that should have been taken to prevent the harm that was suffered. It is usually the failure (i.e. omission to act) on the part of the employer on which the claim will be founded, but it has to be shown that the defendant knew of the risk of injury (by formal complaint) or should have known in all the circumstances. In *Sutherland* v *Hatton* foreseeability was stated in terms of the "particular employee";

> "the threshold question is whether this kind of harm to this particular employee was reasonably foreseeable. The question is not whether psychiatric injury is foreseeable in a person of "ordinary fortitude". The employer's duty is owed to each individual employee not to some as yet unidentified outsider."

Knowledge of the Risk by the Employer: Foreseeability

A further factor is the knowledge of the employer of the risk presented by overwork causing a recognised psychiatric injury. If it is not proved that the defendant employer knew, or ought to have known that psychiatric injury was likely to result, then foreseeability will not be made out. One indicator of the knowledge of the defendant is based on H.S.E. publications, as a result of which employers should have had knowledge of the risks in the early 1990s. These include:

(a) *Mental Distress at Work: First Aid Measures*; HSE, 1992;

(b) *Stress research, Stress Management, putting theory to work*, Prof. Cox, Nottingham University, HSE, 1993

(c) *Guidelines for stress at work*, HSE, 1995.

The Test of Foreseeability

Psychiatric injury is more difficult to foresee than physical injury, but the principle is the same as other cases based on a contractual duty of care (e.g. solicitor-client). If an employer knows that the system of work (including actions of colleagues) may cause mental harm to a fellow employee and the employer does nothing to prevent or control it, then liability will probably be made out. Likewise, there may also be a breach if the employer can foresee that such acts may happen and if they do that harm will result.[11]

In *Sutherland* v *Hatton* a single test was propounded as to whether a harmful reaction to the pressure of the workplace is foreseeable in the context of the individual employee. The test is the same for all types of employment and there are no jobs that are to be considered intrinsically dangerous to health:

(a) The reaction will have two parts:

  (i) an injury to health which is

  (ii) attributable to stress at work.

(b) Whether the injury was foreseeable will depend on the interrelationship between the characteristics of the employee and the requirements of the job, as dictated by the employer.

(c) Expert evidence may assist on this point.[12]

(d) Additional factors have to be taken into account, namely:

  (i) the nature and extent of the work being done by the employee. It will be easier to conclude that harm is foreseeable if the employer is placing unreasonable pressure on the employee. It is also relevant whether other employees are under harmful levels of stress;

  (ii) As to the employee's symptoms, a distinction has to be drawn between symptoms of stress and symptoms of impending breakdown. Stress of itself does not necessarily entail that there will be a breakdown;

11. *Waters* v *Commissioner of Police for the Metropolis* [2000] IRLR 720.
12. *McLoughlin* v *Grovers* [2001] EWCA Civ 1743.

(iii) Harm to health can be foreseeable even in the absence of an express warning from the employee. Relevant factors include, prolonged but uncharacteristic absences from work, which could be stress related;

(iv) Unless the employer knows of some particular problem then the employer is entitled to assume that his employee is up to the job. It is only when there is something specific about the particular job, that the employer has to give greater consideration to the possibility of stress related injury. In general terms the employer can take what he is told at face value;

(v) An employer cannot take the expiry of a GP's certificate as indicating that the employee is now able to return to work and is not at risk of suffering from injury;

(vi) As to how strong the indications have to be:

> "In view of the many difficulties of knowing when and why a particular person will go over the edge from pressure to stress and from stress to injury to health, the indications must be plain enough for any reasonable employer to realise that he should do something about it".

Breach of Duty

The reasonable employer has to take reasonable care to prevent injury to the employees and so the level of care required is relative to:

(a) the foreseeability of harm;

(b) the magnitude of the risk of the harm occurring;

(c) the severity of the harm;

(d) the cost of preventing it;

(e) the justification for running the risk.[13]

13. These are considered in detail in the chapter on duty of care.

Consequently, a breach of duty will be dependant on various factors including the size and scope of the defendant's operation, in particular its resources, in addition to the interests of other employees. Also whether in practice difficulties would arise from re-arranging the work of one employee compared to a substantial part of the remaining workforce. A balance between the interests and requirements of all employees is required, but the employer cannot ignore the need to change the system of work if injury is a reasonably foreseeable consequence of that omission. But an employer is only required to act so as attain and maintain "some good".[14] For his or her part the employee has to balance the benefits of remaining in the employment with the consequent risk to health and employability.[15]

The New Guidelines: A Summary

The Court of Appeal has provided the following summary guidelines that determine the breach of duty of care and causation in stress at work related cases. These form the basis of an assessment of the merits of any work-related stress injury claim:

(1) The ordinary principles of employer's liability apply in this type of case;

(2) The essential test is whether this particular type of harm to this particular employee was reasonably foreseeable, which entails:

  (a) an injury to health and not merely occupational stress; and

  (b) that the harm is attributable to stress at work;

(3) Foreseeability depends on what the employer knows, or ought to have known about the individual employee. Psychological injury is harder to foresee than physical injury, but it may be more easily foreseen in an individual employee. An employer is entitled to assume that an employee is able to withstand the normal pressures of the job in the absence of known particular pressures;

14. Per Hale LJ.
15. On the duty to dismiss an employee see *Coxhall* v *Goodyear Great Britain Limited* [2002] EWCA Civ 1010: it is an issue of the magnitude of risk of a known injury.

(4) The test is the same irrespective of the employment;

(5) Factors relevant to the threshold issue include:

(a) The nature and extent of the work done by the employee. Whether the workload is excessively heavy or whether the intellectual demands are above that which the employee could be expected to achieve;

(b) Signs from the employee of impending harm to health. This includes knowledge of whether the employee has a particular vulnerability. If the employee has suffered from this type of illness in the past then this is a good indicator that the employer could be approaching the threshold criteria if no action is taken;

(6) Employers are entitled to take what the employee says about his or her condition at face value, unless there is good reason to the contrary. There is generally no need to make searching enquiries of the employee;

(7) The trigger on the duty to take action on stress at work must be related to clear signs of impending harm to the employee;

(8) A breach of the duty of care is based on a failure to take steps which are reasonable in the circumstances, taking account of the severity of the risk of harm, the gravity of the harm that may occur, the costs of practically preventing it and the justification for running the risk;

(9) In deciding what is reasonable for the employee one has to take into account the resources and demands of the employer, and this includes the interests of other employees;

(10) An employer is only under a duty to take steps that will do some good and this may be a matter for expert evidence;

(11) If an employer offers a confidential advice service then there is unlikely to be a breach of duty;

(12) If the only reasonable and effective step would have been to dismiss or demote the employee, the employer will not be in breach in allowing the employee to continue in the job;[16]

16. But see *Coxhall* v *Goodyear Great Britain Limited* [*supra*].

(13) In all cases it is necessary to identify the steps which the employer could and should have taken before finding the employer in breach of the duty of care;

(14) The employee has to prove that the breach of duty caused or materially contributed to the injury suffered. It is not enough to show that mere occupational stress has caused the harm;

(15) If the injury suffered has more than one cause the employer should only pay for that portion of the harm that is attributable to the breach of duty of care. The defendant has to raise the issue of apportionment;

(16) An assessment of the damages will take account of any pre-existing disorder or vulnerability and the chance that the claimant would have succumbed to a stress related disorder in any event.

Causation

A claimant must prove the breach of duty caused the harm, as opposed to being merely associated with it. Where there is a number of causes then causation can be problematic.[17] Furthermore, where a claimant has a vulnerable personality, of which the employer did not know then this can present a significant problem for the claimant on causation. These are relevant factors. However, a claimant does not have to prove that the breach of duty caused all his injury. The starting point is to prove that the negligence made a material contribution to the injury.[18] Once that is proved then an exercise in apportionment can be undertaken.[19]

The Approach to Causation in Psychiatric Injury Cases

In summary, *Holtby* v *Brigham and Cowan (Hull) Limited*[20] sets down the principles on apportionment between the various causes (save those that are *de minimis*). A defendant is liable only for that injury which it has caused. This presupposes that the injury is "divisible" as to its cause, which is substantially a

17. See *Wilsher* v *Essex Health Authority* [1988] AC 1074. [1988] 2 WLR 557. But see also *Fairchild* v *Glenhaven* [*supra*] on this point.
18. As in *Bonnington Castings Limited* v *Wardlaw* [1956] AC 613.
19. This issue is fully discussed in Chapter 2 on causation.
20. [2000] PIQR Q293. [2000] 3 All ER 421.

matter of medical evidence. In *Sutherland* v *Hatton* the problem was approached in this way:

(1) If the harm is truly indivisible then a tortfeasor who has made a material contribution is liable for the whole loss. The usual example given to define the "indivisible injury" are two defendants struggling for the same gun, who kill the claimant. There is no way in which the liability can be apportioned and so they are each liable for all of it. In many cases psychiatric injury based on work related stress will be indivisible in that sense, but there may be other pre-existing constitutional factors which are relevant to causation;[21]

(2) Where the harm is divisible then an apportionment should be carried out:

> "Where it is possible to identify the extent of contribution that the defendant's wrong made to the claimant's damage then the defendant is only liable to that extent and no more."[22]

(3) Complications arise in the following circumstances:

 (a) Where in the case of post traumatic stress disorder the condition is over-come by a grief reaction;[23]

 (b) Where the whole of the psychiatric injury was caused by the witnessing of the death of a third party;[24]

(4) In the case of an exacerbation of a pre-existing condition or the accelerated effect on a pre-existing vulnerability, then the general damages award will reflect only the exacerbation or acceleration. In addition, the financial losses need to take into account some of the contingencies, namely that the claimant would have suffered a stress related illness in any event. The multiplier can be reduced to take account of these factors. For example, in *Page* v *Smith*[25] the future loss of earnings claim of a teacher was reduced on the multiplier

---

21. The approach in *Vernon* v *Bosley [No.2]* [1997] 1 All ER 614, where there was no apportionment between tortious and non-tortious causes does not seem to have been followed in recent authority, but the point was not expressly dealt with by the Court of Appeal.
22. *Clerk and Lindsell on Torts* 18th Edition [2000] para 2–21.
23. *Calascione* v *Dixon* [1993] 19 BMLR 97.
24. *Vernon* v *Bosley (No.1)* [1997] 1 All ER 577.
25. [*supra*].

from 10 to 6 to reflect the possibility of future complications arising from the employment.

Application of the Principles: Case Studies

Sutherland v Hatton: (Mrs. Hatton)

Mrs. Hatton had been a teacher since 1976. In October 1995 she was signed off work due to depression and she retired on health grounds in 1996. At trial her claim was successful and the defendant (School Governors) appealed. It was found that her workload at the school was no greater than any other teacher in a similar school and she had made no complaint about it. When the school changed its curriculum no other teacher found this a particularly excessive strain. Whilst she was off work in 1993 she did not tell anyone about her condition and nor did she attribute it to overwork. Her pattern of illness and absence were not apparently stress related. Consequently, the claim failed on the threshold of attributability, but there was also a problem on causation. This was a case where no-one could be blamed for the injury that curtailed her career. The defendant's appeal was allowed.

Sutherland v Hatton: (Mrs. Jones)

Mrs. Jones was employed as an administrative assistant at a local authority training centre from 1992 to 1995. In 1995 she was absent from work with anxiety and depression and she never returned and was made redundant in 1996. The claimant was successful at trial and the defendant employer appealed. The claimant's professional tasks varied and there was constant time pressure on her. She worked in excess of her 37 hour contractual week and the defendant accepted that it was a known gamble at the time to have let her take on all the responsibilities. She complained of overwork to her line manager at an early stage and that she had been threatened with non-renewal of her post if she continued to complain. Although extra assistance was earmarked for her, this never materialised. In 1994 the claimant invoked the formal grievance procedure. This was not heard until after she had left her employment on health grounds and overall she was treated unreasonably as to the reaction to her complaints of overwork. In any event, no assistance was provided to her when the defendant knew she required assistance as the demands on her were unreasonable.

In dismissing the defendant employer's appeal the Court of Appeal found that:

(a) The issue was whether the harm to the claimant was reasonably foreseeable, in that the actions of the employer were placing unreasonable demands on her and this was likely to cause her injury. It dismissed the appeal and emphasised that the case was based on an injury caused by the maltreatment of one employee by another, of which the employer knew but failed to act upon or prevent;[26]

(b) The fact of the formal complaint based on her alleged unreasonable treatment rendered the injury foreseeable. The injury could thereafter have been prevented and this omission caused her injury.

Sutherland v Hatton (Mr Bishop)

Mr. Bishop worked from 1979 to 1997 for a company that undertook mixer cleaning and graphite blowing. In 1992 his company was taken over by an American company and he was required to do a greater variety of tasks. Most employees welcomed the new working conditions, but Mr. Bishop did not. He requested to go back to his old job, but that would have led to an impractical re-organisation. It was held there was nothing unreasonable about the demands of the job, and the reality was that Mr. Bishop was unable to deal with the changes. He had been to his doctor who advised him to change jobs but the employers were not told of this fact. Even if they had been informed there was nothing they could reasonably have been expected to do. The employer could not be in breach of duty of care where it failed to dismiss someone who indicated a desire to continue working.[27] The appeal of the defendant employer was allowed.

26. See also *Waters* v *Commissioner of Police of the Metropolis* [2000] IRLR 720.

27. But see *Coxhall* v *Goodyear Great Britain Limited* [*supra*].

Sutherland v Hatton (Mr. Barber)

Mr. Barber was employed as a school teacher from 1984 until 1996 when he ceased work on medical advice and retired early in 1997. He succeeded at trial and the defendant employer appealed. In 1996 the school at which he worked was a comprehensive under pressure, and as a head of department the claimant became an "area of expertise co-ordinator" in mathematics. His combined duties entailed long hours and many of his colleagues were suffering from overwork at this time, but he was no more overworked than his colleagues. The claimant became depressed in 1995, but did not inform the school. There was a subsequent meeting where he told the headteacher of his concerns. All in all it was found he had not informed the school sufficiently of his difficulties such that it was incumbent on it to take steps to prevent him suffering injury. The appeal of the employer was allowed.

Post Sutherland v Hatton Case Law

In *Young* v *Post Office*[28] the claimant was a transport technician who was promoted to workshop manager in 1988. The role was a stressful one and included managing a group of four employees. He was not provided with formal training in the operation of a new computer system. By 1994 the claimant was showing clear signs of stress and was taking anti-depressants. He suffered a nervous breakdown and had four months off work and then returned to work in September 1997 at 80% recovery, but supposedly on a gradual reintroduction. He was unable to cope and had to leave.

At trial it was found that the defendant knew that the claimant needed to be looked after and that the defendant had created a rehabilitation package that it did not follow through. The employer sought to appeal on the basis that it had acted reasonably and that the injury was not foreseeable. The Court of Appeal dismissed the appeal and held that:

(1) As the claimant had already suffered from stress, the recurrence was plainly foreseeable once he returned to work;

28. [2002] ECWA Civ 661. See also, *Foumeny* v *University of Leeds* [2003] EWCA Civ 557.

(2) The defendant had some justification to assert that the claimant had been reintroduced to work on a flexible basis, but it had failed to carry it through;

(3) The claimant was not negligent in causing his own injury. He always tried to work hard at what he was required to do but he was psychiatrically vulnerable. It would be a rare case where the claimant would have to bear responsibility for the injury caused by working too hard.

In *Pratley* v *Surrey County Council,*[29] the claimant had worked for the defendant council since 1986 as a case manager. It was accepted by the defendant that the job created excessive pressures on the claimant and there was a risk of stress that was injurious to health.

In dismissing the claim reference was made to *Sutherland* v *Hatton*. Applying the principles in that case, it was held that the fact the defendant was aware and conceded pressure and stress could cause stress related injury was insufficient. There was a crucial difference between general knowledge that stress can cause illness and knowledge that a real risk has arisen in a particular case. In this case there was nothing to alert the defendant of the claimant's stress related illness.

Contributory Negligence

It is arguable in some cases that a claimant has brought injury upon himself or herself. In *Sutherland* v *Hatton* it was held that if there is no knowledge (actual or constructive) of the defendant employer then it is unlikely that the injury was reasonably foreseeable. Nevertheless the issue remains a matter of degree. An offer by the claimant to continue to work does not in itself amount to contributory negligence in ordinary circumstances. But, in contrast, where a claimant employee fails to make a material disclosure to the employer about his or her medical condition (and specifically fails to make proper complaints) then the employer could not be expected to anticipate such a risk.[30] There is a crucial distinction between a general knowledge that stress could cause illness and a known real risk in a particular case.

29. [2002] EWHC 1608.

30. *Pratley* v *Surrey County Council* [2002] EWHC 1608. For more on this topic, see *Contributory Negligence* EMIS.

Quantum of Damages

In the context of work related stress claims the particular factors to take into account on general damages will include:

(1) Age of the claimant;

(2) Whether and to what extent the injury (for example, a mental breakdown) was due wholly or substantially to the pressures of overwork;

(3) Whether there are particular exacerbating aspects of the psychiatric injury, for example suicide attempts, or a wider psychiatric disorder (e.g. agoraphobia);

(4) Whether the claimant will be fit to return to work in the future and if so when;

(5) If a return to work is envisaged will it be in the previous pre-accident job. The loss of ability to undertake pre-accident employment may form the basis of a loss of congenial employment claim.

Further Guidance

Case Law

The schedule on page 222 contains recent case law that provides some practical guidance as to the overall assessment of damages. There are particular points of note.

- As to the future loss of earnings claim: if the claimant is unable to return to the pre-accident job then consider the residual earning capacity.

- The claimant has to provide evidence of the residual earning capacity.

- A central issue will be whether a handicap in the labour market award should be made as opposed to a future loss award based on a multiplier/multiplicand assessment.

- If the claimant is likely not to have worked until retirement in any event and/or there was a pre-existing condition, then the multiplier can be reduced to reflect this contingency. The higher the risk the greater the reduction.

Case Name	Quantum	Comment
Sutherland v *Hatton* [2002] IRLR 263. [2002] 2 All ER 1.	Total award: £90675 £6,000 General damages. £46,876 Past loss (gross) £53,560 Future loss.	• Claim dismissed on appeal, but assessment was made of the judge's findings at para 132. • The Court of Appeal held that as C did not adduce evidence of her earning capacity then a handicap in the labour market award should have been made. An award of £10,000 would have been made.
Somerset County Council v *Barber* (considered in *Sutherland* v *Hatton*).	Judgment: £101,041 General damages £10,000 (for moderately severe psychiatric illness). • Future loss award made.	• C was fit to return to work. • A multiplier of 4 was taken in relation to his retirement age. • A discount on the multiplier was needed for the possibility that C would not have worked until retirement age in any event. The judge was wrong not to reduce the multiplier. Multiplier was reduced to 1.
Sandwell MBC v *Jones* (considered in *Sutherland* v *Hatton*).	Judgment £157,541 General damages £22,000 Future loss of earnings and pension: £94,820.	• Born in 1953: acute anxiety that developed into agoraphobia accompanied by mild depression with obsessive compulsive symptoms. • Claim dismissed on appeal on liability.

Case Name	Quantum	Comment
Baker Refractories Limited v *Bishop* [considered in *Sutherland* v *Hatton*].	General damages: £7,000.	• Mental breakdown with suicide attempt. • Future loss adjourned because of potential intervening back injury.
Witham v *Hastings and Rother NHS Trust* [2001]. QBD (Unreported) 4.10.01	Total award: £140,000 General damages: £17,500 Loss of status as grade G nurse: Congenial employment £7500.	• Female, 37, subjected to excessive hours as a nurse. • Obvious deteriorating condition. Depression and off work for about one year intermittently. Treated in clinics. • Retired on health grounds due to depressive illness. Remained on medication. No full recovery, but there would be some recovery.
Rowntree v *Commissioner of Police for the Metropolis* [2001] QBD Unreported 26.10.01	• Total award £135,236 • General Damages £15,000 • Loss of congenial employment: £6,000 • Loss of future pension £25,000	• 38 year old female. Employed in Child Protection Team and suffered moderate PTSD • Off work for 2 months from Xmas 1995, but suffered 2 breakdowns. Retired on medical grounds in 1998. • Condition would improve and could work in the future.

Case Name	Quantum	Comment
Young v *Post Office* [2002] ECWA Civ 661	Total award £94,000. No further breakdown provided.	• C was a personal transport technician. • Overwork caused a nervous breakdown and 4 months off work. • Returned to work for a short period but then could not cope and was dismissed.
Cowley v *Mersey Regional Ambulance Service NHS Trust* [2001] QBD Unreported 16.2.01	Total award £139,331. General damages: £15,000 Future loss of earnings: £82,554 based on six years' loss. £12,082 loss of pension rights.	• Illness caused by excessive workload. • Claimant suffered two breakdowns and second was serious and required electroconvulsive therapy.

Summary of this Chapter

Contained below is a summary of this chapter that also highlights points from the defendant and claimant practitioners' perspective:

Summary of Issue	Claimant's Perspective	Defendant's Perspective
The main case law: *Sutherland* v *Hatton* All case law must be read on the basis of this case.	This case provides the rubric for the proper approach to this type of case.	As with claimant.
Duty of care The duty of care must not be set too high or low. The fact it is a "recognised" psychiatric injury is not relevant.	Initial investigations: consider if it a recognised psychiatric injury. Medical evidence is required and other investigations are dependent on this point.	Obtain medical evidence on whether it is a recognised psychiatric injury.
Duty of care/Foreseeability Foreseeability of injury is the primary factor on liability. Was the risk of injury foreseeable in this *particular* employee?	Assess foreseeability in the "particular" employee. Evidence of particular known (and unknown traits) is required. The personnel file must be disclosed and specific characteristics noted.	Assess the "particular" employee. Evidence from medical records, Benefits Agency, personnel file is required. Can it be shown there was no foreseeability on this basis.
Foreseeability Knowledge of the defendant of the risk of injury is an important factor.	Proof that the type of injury was known by the defendant. Publications (HSE) assist in proving knowledge.	Assess the date of knowledge of risk of injury. Consider HSE publications (and other expert sources). Obtain instructions from senior executives and managers.

Summary of Issue	Claimant's Perspective	Defendant's Perspective
Liability: Breach of Duty Foreseeability of injury depends on the interrelationship between the characteristics of the employee and the requirements of the job as required by the employer.	Obtain full instructions on the basis of the complaint and an analysis of system of work. When was the defendant put on notice? Assess other colleagues' ability to cope. Consider the symptoms against the timing of the breach. Medical notes must be analysed. Expert evidence may be required.	Assess the system of work and identify the basis of the complaint. Obtain statements from managers on the system of work. Did the claimant complain? Could colleagues cope with system of work? Analyse medical notes and identify date of presentation of symptoms. Does this tally with the complaint? Expert evidence is required.
The Court of Appeal A summary of the 16 guidelines on liability and causation.	Go through this checklist to ensure that all evidence is obtained on each point and to assess the merits of the claim.	Same as claimant: use this checklist as the rubric to defend the claim. Consider contributory negligence.
Causation/Damages: Causation will usually be approached on an apportionment basis, as long as a material contribution to the injury is proved as being caused by the breach. *Holtby* v *Brigham and Cowan Limited* provides the principles.	Medical evidence must cover causation. Did the breach of duty cause the symptoms and if so, why? Consider pre-existing or other factors on a % basis. Apportionment is usually required. What is the ability to work in the future and handicap in the labour market?	Medical evidence must deal with causation. Percentage apportionment is usually required. Is the claimant able to go back to the same type of work? Would he have given up work anyway? Assess residual earning capacity and/or handicap in the labour market?

Section 2: The Medical Perspective

Overview

The Trailer to the 1990 Labour Force Survey (HSE, 1990), indicated that mental illness caused or aggravated by work, including stress, anxiety or depression, was the second most frequently reported cause of occupational illness after musculo-skeletal disorders. It has been estimated that up to 30% of all sick leave is related to such conditions.

In addition to the immediate psychological effects upon the employee, surveys performed by the CBI and Department of Health suggest that there is a huge annual cost to employers, through absenteeism and lost productivity.

Stress may also play a role in workplace accidents, and contribute to other health problems, such as high blood pressure, heart disease, alcohol and drug abuse, and skin disorders.

The Bodily Reaction to Stress

"Healthy" stress

A moderate amount of stress due to the normal challenges of the workplace is generally perceived to be beneficial. The individual will feel motivated and able to cope. This is sometimes referred to as the zone of the "three C's": cool, calm, collected.

Fight or flight

With rising levels of stress, the nervous system reacts by stimulating the release of hormones such as adrenaline, stored in the adrenal glands adjacent to the kidneys. The nerves supplying the heart also have direct effects on the circulation. These changes prepare the body for action – literally "fight or flight".

The effects are noticeable in the circulation and lungs, for example raised heart rate, respiratory rate and blood pressure. Dry mouth, increased sweating and palpitations are common. There will often be a subjective sensation of anxiety.

Exhaustion and burnout

In the short term, the fight or flight reaction is a normal physiological adaptation to stress. However, the reaction cannot be sustained indefinitely. If coping strategies are ineffective, then bodily reserves will eventually become depleted, leading to physical and mental exhaustion. The end stage is burnout, and frank physical and mental illness.

The physical symptoms may include:

(a) Dizziness and palpitations

(b) Hyperventilation

(c) Raised blood pressure

(d) Non-specific back, chest or abdominal pains

(e) Disturbances of digestion or bowel habit

(f) Tension headaches or migraine

(g) Aggravation of asthma

(h) Aggravation of skin disorders such as eczema.

The mental symptoms may include:

(a) Irritability

(b) Anxiety and panic attacks

(c) Bouts of uncontrollable crying

(d) Depression

(e) Low libido

(f) Loss of interest in relationships, hobbies ("anhedonia") and personal appearance

(g) Impaired concentration and work performance

(h) Abnormal sleep patterns

(i) Abnormal eating patterns leading to weight gain or loss

(j) Misuse of cigarettes, alcohol and prescription drugs such as tranquillisers and sleeping pills.

Risk Factors: Stress Triggers

Physical stressors

Physical stressors tend to be more important in the factory setting. These may include:

(a) Extremes of temperature

(b) Noise

(c) Vibration

(d) Exposure to fumes, chemical, biological or radiation hazards

(e) Physical fatigue due to excessive working hours or irregular patterns of shift work

(f) Poor lighting.

Mental stressors

Many mental stressors are capable of contributing to stress-related illnesses. For example, the stress of promotion may be just as stressful as anger and resentment following a demotion. Monotonous, unfulfilling work at a junior level may be subjectively just as stressful to that employee, as the demands placed on a senior executive. Other examples include:

(a) Conflict with subordinates, colleagues or managers

(b) Competition with colleagues

(c) Uncertainty due to inconsistent instructions, poor communication or lack of clear objectives

(d) Inflexible work schedules

(e) Company restructuring or mergers, particularly where workloads increase due to "downsizing" of the workforce

(f) Fear of sanctions if productivity targets are not met

(g) Loss of support systems, for example isolation from sympathetic colleagues through extensive travel or working from home

(h) A "macho" work culture that permits no sign of weakness.

Individual Factors

Individual factors may also play a role in determining how a particular worker copes or fails to cope with work-related stress. There is great individual variation. These may be external factors, such as lifestyle, family relationships and other sources of social support; or internal constitutional factors, such as personality, or pre-existing physical and mental health problems.

Sometimes loss of social support, for example that following divorce or bereavement, may be sufficient to precipitate an inability to cope with work-related stresses, even where the demands of the job are unchanged. Lifestyle factors such as patterns of eating, drinking, sleeping and exercise may be relevant.

Personality factors may affect self-confidence, organisational skills, and the ability to react to unpredictable or difficulty situations. For example, rigid, obsessional personalities may experience greater stress in posts that require flexibility and rapid decision-making.

Post-Traumatic Stress Disorder

Post-traumatic stress disorder ("PTSD") is a specific type of anxiety disorder which usually follows a discrete, shocking event, such as witnessing a fatal accident. The incidence tends to be greater among service personnel, hospital and emergency services. Bank, post office and security staff may also be exposed to violent criminal incidents involving threats and intimidation, falling short of physical injury. The typical symptoms include:

(a) Recurrent and intrusive memories ("flashbacks")

(b) Avoidance of any situation that resembles the original incident

(c) A state of abnormally increased responsiveness to sensory stimuli ("hypervigilance")

(d) Nightmares

(e) Insomnia

(f) Irritability

(g) Difficulty concentrating.

There is often overlap with other psychiatric disorders, such as anxiety, depression, obsessive compulsive disorders, and abuse of alcohol and other substances.

Risk Management

Information

Employees should be provided with the information which allows them to recognise when something is wrong, and what to do, so that they can get help for themselves or colleagues. This may take the form of stress awareness and stress management courses.

Assessing risk

Risk assessment may be undertaken at a company level through employee surveys. These may help to identify areas of particular concern, such as a department that is showing abnormally high levels of absence, or complaints of bullying or harassment from line managers.

At the individual level, pre-employment assessments are helpful to identify the competence of an individual for a particular post, and their training needs.

Management of Stress at Work

Many employers do not have occupational health departments. Where they do exist, they may have an important role to play in identifying sources of stress in the workplace, identifying other affected employees, and liasing with managers to implement remedial action.

Other measures may be adopted including:

(a) *Counselling*. Many employers provide access to confidential counselling services for stress and a wide range of other problems, such as bereavement, debt and alcohol abuse.

(b) *Relaxation therapy*, using self-help techniques such as relaxation tapes and meditation.

(c) *Psychological or psychiatric treatment*. In more serious cases, formal treatment may be required. This will require a careful re-appraisal of fitness to work upon return to the workplace. A staged return, through initial part-time work, may be required, together with action to reduce the relevant stressors to the lowest possible level. In some cases rehabilitation may prove impossible, in which case cessation of employment may have to be considered, in order to prevent relapse.

Further Reading

Teasdale, E, Creed, F. "Health and Mental Illness at Work". In: *Hunter's Diseases of Occupations* (9th edition). London, Arnold, 2000.

Von Onciul, J. "Stress at Work". In: *ABC of Work Related Disorders*. London, BMJ Publishing Group, 1997.

CHAPTER 11 OCCUPATIONAL ASTHMA AND INDUSTRIAL DERMATITIS

SECTION 1: THE LEGAL PERSPECTIVE

Occupational Asthma: Overview

Occupational asthma results from narrowing and inflammation of the airways which causes breathlessness, wheezing and coughing. Further details are set out in the medical section of this chapter. With industrial activity it is particularly common in those workers who are exposed to "sensitisers" in the course of their employment.[1] Later in this chapter (at page 246), further consideration will be given to occupational dermatitis, as in practice this injury occurs in similar industrial circumstances to occupational asthma (i.e. contact with hazardous substances) and in some cases injury is caused from the same source (e.g. latex exposure). Accordingly, the legal principles set out in section 1 will apply broadly to cases of industrial dermatitis as well.

The Duty of Care

The Date of Knowledge of the Risk of Injury

The duty of care imposed on an employer to an employee is circumscribed according to whether the employer knew (actually or constructively)[2] of the risk of causing asthma caused by the industrial process in question.[3] To fix a date when

1. Note that to claim social security benefit in relation to asthma a claimant must prove he worked with an employer for 10 years and has a 14% disability.
2. On constructive knowledge see *Ogden* v *Airedale Health Authority* [1996] 7 Med LR 153 at 159.
3. The duty of the employer is set out in Chapter 1. See in particular *Stokes* v *Guest, Keen and Nettlefold (Nuts and Bolts) Limited* [1968] 1 WLR 1776.

employers ought to have known of the risk of injury the following publications and regulations provide guidance:

(1) *Toxic Substances in Factory Atmospheres* [Ministry of Labour 1960]

(2) Section 6 of the Health and Safety at Work Act 1974 imposes a statutory duty on manufacturers to supply relevant information on these substances.

(3) The Social Security (Industrial Injuries Prescribed Diseases) Regulations 1985.

(4) *Preventing Asthma at Work-How to Control Respiratory Sensitisers* [HSE 1994]

(5) *Health and Safety at Work; Occupational Health* [1994 HSE]

(6) *Preventing Asthma at Work* [1994 HSE]

(7) *Medical Aspects of Occupational Asthma* [HSE 1998]

Employers must obtain and maintain a level of knowledge of the risks of workplace injury to the standard of the reasonably competent employer.[4] Factors such as knowledge of the gravity of the risk of injury and general industry standards are relevant considerations. However ignorance of a known risk of injury will not generally afford a defence to a claim. Where knowledge regarding the causes and effects of occupational asthma has increased over time, an employer is expected to keep pace with any significant developments. However a reasonable employer would still be entitled to an acceptable period in which to change working practices.[5] In practice, it will be difficult for an employer to argue that some injury was not reasonably foreseeable, in cases where an employee was exposed to a known respiratory sensitiser.[6]

4. See Chapter 1 where this is discussed in detail.
5. As to the case law on this issue see Chapter 1.
6. As to the issue of knowledge of the defendant employer see *Ogden* v *Airedale Health Authority* [1996] 7 Med LR 153 at 158. Note that reasonable foreseeability would not be a relevant consideration in relation to any statutory duty, unless expressly provided.

Breach of Duty of Care

Case Law

Assuming it is proved that a defendant employer should have known of the risk of injury presented by respiratory sensitisers in the workplace, then the claimant has to prove a breach of duty. Proof that some injury was foreseeable is sufficient to found liability.[7] In *Ogden* v *Airedale Health Authority*,[8] a radiographer was employed in a hospital from 1979 to 1991. He initially complained of dry eyes and a sore throat. He also claimed the symptoms were caused by the chemicals he used at work, particularly as he noted an improvement in his symptoms when he was away from work.

In 1987 the claimant was admitted to hospital with a diagnosis of atypical pneumonia. He was off work for five and half months. Having returned to work in 1988 he suffered breathing difficulties, and he was again admitted to hospital. By 1991 it was found that his symptoms were probably caused by exposure to x-ray chemicals and he took ill-health retirement. The claimant alleged x-ray chemicals had caused occupational asthma. In summary the allegations of negligence were:

(a) a failure to provide a safe working environment free from fumes;

(b) a failure to monitor and ventilate the premises;

(c) an omission to warn him of the risks of injury, or to provide proper protective equipment;

(d) Reliance was placed on the Control of Substances Hazardous to Health Regulations 1988.[9]

7. *Hughes* v *Lord Advocate* [1963] AC 837.
8. [*Supra*].
9. These regulations are discussed below.

The Findings at Trial

The claimant succeeded on liability for the following reasons:

(a) *There were risks of injury* caused by the system of work and the judge held that:

> "I am satisfied.... that during the time of the plaintiff's employ... there were in use in his place of work chemicals which were capable of having an irritant effect upon the skin of the employees and some of which were sensitisers capable or producing skin allergy and asthma and that during some of the activities which the staff were required to carry out such as a mixing of chemicals there was a likelihood of recommended exposure limits being exceeded."[10]

(b) *The knowledge of the defendant*

The defendant was well aware that staff were complaining of health problems due to hazardous chemicals. It had produced a list of chemical products used in the x-ray department in 1990. As to the period prior to 1990 reliance was placed on the fact that bottles and containers of chemicals had warnings and other relevant information. Moreover, the defendant could have obtained such information from the manufacturer. Finally, it was put on notice as the claimant was not the only person presenting with symptoms.

(c) *Precautions that should have been taken*

There was a duty to keep the fumes as low as possible and so control them at source. This should have been achieved by an extractor. Spillages should have been avoided and masks should have been provided if there was a risk of fumes.[11]

(d) *Precautions that were in fact taken*

The defendant had paid scant regard to this aspect of health and safety prior to 1988. There was no ventilation and no procedures for the safe handling of chemicals. Staff had not been provided with the appropriate protective equipment and there was no monitoring of the level of fumes. Some

---

10. At 158.
11. These were described as "common sense" by the expert.

improvements were made post 1988 when the COSHH regulations were implemented, with the introduction of respirators to replace masks. Generally there was a better approach to the level of fumes but there were still shortcomings in relation to the ventilation system. The judge found that:

> "The failure to protect the plaintiff against exposure incidental to hand mixing by providing local exhaust ventilation, warnings as to the dangers and/or protective equipment was in my judgment equally plainly negligent."[12]

Further Case Law: Moving or Dismissing the Employee

Coxhall v *Goodyear Great Britain Limited.*[13] The claimant successfully alleged at trial that he was injured in the course of his employment as a paint sprayer. The defendant appealed. The claimant was provided with protective goggles and a mask in addition to a respirator. These measures were all held to be reasonable. However, the claimant suffered from an unknown constitutional asthmatic condition. In April 1996 the claimant saw the works' doctor complaining of headaches and giddiness. It was recommended that he should not work with the paint in the future. This information failed to reach the manager and the claimant remained at work on the line using paint. In May 1996 the claimant fell unconscious and was formally diagnosed as suffering from occupational asthma.

The defendant accepted that if the recommendation of the doctor had been known then the claimant would have been dismissed, but it was not acted upon. Nevertheless, the defendant denied it was under a duty to remove the claimant from his job on the basis that he was unsuited to it. It was a matter for the employee to decide whether or not to run the risk of continuing on with the job.[14] The case law was conflicting, but the Court of Appeal held that the principle established in *Withers* v *Perry Chain* was now dependant on the magnitude of the risk in question. Employees may have to be dismissed for their own good based on the magnitude of risk that they would run by remaining in the job. Here the claimant should have

12. At p161.
13. [2002] EWCA Civ 1010.
14. See *Withers* v *Perry Chain Co. Limited* [1961] 1 WLR 1314, *Henderson* v *Wakefield Shirt Company Limited* [1997] PIQR 413, where it was held there was no duty to dismiss an employee in this situation.

been dismissed as it was no longer safe for him to continue in that type of work.

Statutory Duty

Overview

Prior to the Control of Substances Hazardous to Health Regulations ("COSHH"), a claim was based on common law negligence in addition to relevant industrial legislation (for example the Factories Act 1961, s.63).[15] Additional statutory regulation was provided by the Workplace (Health, Safety and Welfare) Regulations 1992[16] and the Personal Protective Equipment at Work Regulations 1992/1999. But the principal regulations are COSHH regulations. These regulations will probably now cover most claims that have arisen over the past twelve years:

(1) COSHH Regulations 1988[17]

(2) COSHH Regulations 1994[18]

(3) COSHH Regulations 1999[19]

(4) COSHH Regulations 2002.[20]

COSHH regulations have previously imposed an absolute duty on the employer to comply with the regulations.[21] In *Williams* v *Farne Salmon & Trout Ltd*[22] it was held that the COSHH Regulations 1988 imposed an absolute duty on employers, which was not qualified by any provisions relating to knowledge or foreseeability. In *Bilton* v *Fastnet Highlands Ltd*[23] the claimant claimed injury caused by exposure to respirable prawn protein and other substances which caused her to develop occupational asthma. These were hazardous substances under COSHH Regulations 1988. It was held that once the

15. This applies to exposure before 1989.
16. E.g. Reg 6 that requires effective ventilation.
17. Effective from 1.10.89.
18. Effective from 16.1.95.
19. Repealed by the 2002 Regulations.
20. Effective from 21.11.02.
21. Note that the 2002 Regulations seem to impose a "reasonable practicability" test: this may undermine the absolute duty.
22. [1998] SLT 1329.
23. [1998] SLT 1323.

statutory regulations were relied upon, the onus of proof that the defendant had complied with the requirements of the regulations was transferred to the defendant.

COSHH 1994 and 2002

The new COSHH Regulations 2002 have repealed the COSHH Regulations 1999. To provide a comparison between the COSHH Regulations 1994 and COSHH 2002, a schedule of the particulars regulations is set out below.

The Control of Substances Hazardous to Health Regulations 1994[24]	The Control of Substances Hazardous to Health Regulations 2002
Reg. 3: the duty is imposed on the employer so far as "reasonably practicable". The regulations apply to self employed persons. "Substance" includes solids or vapours and "substance hazardous to health" means any substance in part 1 of the approved supply list or in Schedule 1.	Reg. 3: a duty is imposed on the employer (which includes the self-employed) "so far as reasonably practicable" to persons who may be affected by the employer's work. With some exceptions this covers persons who are not employees.
Reg. 6: Requires the employer to carry out a risk assessment as to the exposure to hazardous substances. That assessment has to be reviewed.	Reg. 4: sets out the prohibitions relating to certain substances.
Reg. 7: The employer must prevent or properly control the exposure of his employees to hazardous substances. This includes the provision of protective equipment.	Reg. 6: Employers must assess the risk to health created by exposure to hazardous substances.
Regs 8 and 9: The employer must ensure that all protective equipment is used and applied properly (as provided under Reg. 7). It must also be properly maintained, kept in good working order and regularly tested.	Reg. 7: Employers must ensure that exposure of employees to hazardous substances is prevented, or where this is not reasonably practicable adequately controlled. This includes substitution of less hazardous substances where possible. Further specific practical requirements are given for the safe handling, storage and transport of the hazardous substances.[25]
Reg. 10: The employer must monitor the employee's exposure in the workplace.	Regs. 8 and 9: Provide for the proper use of control measures and to ensure that any control measure is maintained in an efficient state.
Reg. 11: A system of health monitoring must be undertaken by the employer.	Reg. 10: requires that the exposure to hazardous substances is monitored.

24. The 1994 Regulations are cited as they are probably most relevant to cases arising in the last 10 years.
25. Note that if protective equipment is provided the employer must ensure it is worn.

The Control of Substances Hazardous to Health Regulations 1994	The Control of Substances Hazardous to Health Regulations 2002
Reg. 12: employers must ensure that employees are given suitable and sufficient instruction, training and information on hazardous substances such that they know the risks to health caused by exposure.	Reg. 11: Employers should, where appropriate, monitor the health of employees.
	Reg. 12: Suitable information, instruction, and training must be given to employees, including details of the substances hazardous to health.
	Reg. 16: exempts application of these regulations to the Ministry of Defence.
	Schedules 1 and 2: set out the definition of carcinogens and the list of prohibited hazardous substances.

Causation

Overview

Practitioners need to assess at an early stage whether a claimant is suffering from an occupational asthma.[26] Asthma is common in the working population and so causation is a central issue. Supportive expert medical evidence is essential, before extensive investigations into liability are conducted. If the medical evidence cannot definitively state on the balance of probabilities that asthma has been caused or contributed to by the employer's negligence then the claim will fail.[27]

26. Occupational asthma is a prescribed disease: see also Social Security (Industrial Injuries) (Prescribed Diseases) Regulations 1985.
27. Conversely, defendant practitioners need to address this point which may save the costs of further investigations.

Recent Legal Developments

Causation is a critical aspect of any case, as a claimant must prove the causal connection between the injury and system of work. The legal principles are set out in detail in the chapter on causation to which reference should be made. But in summary the proper approach to causation in industrial asthma cases should be as follows:

(a) The approach taken in *Bonnington Castings Limited* v *Wardlaw.*[28] A claimant has to prove on balance that the negligence caused or materially contributed to the injury, although any contribution which was not de minimis was "material[29]". Hence proof of a "material contribution" on the medical evidence is the starting point, but this of itself does not entitle the claimant to full compensation;

(b) Further, *Holtby* v *Brigham and Cowan Ltd,*[30] held that the defendant should still only be liable for the amount of the contribution to the injury in the case of a divisible injury (which includes asthma);

(c) Pre-existing conditions of the claimant which are exacerbated or accelerated by the negligence will be taken into account on causation and medical evidence must be obtained on the issue.

A Practical Application of Causation

In *Ogden* v *Airedale Health Authority* the medical experts were divided on the issue of causation. The defendant's expert concluded there was insufficient evidence to prove occupational asthma. He ascribed the claimant's symptoms to panic attacks, but if that was rejected then he was "stuck for an explanation". The judge rejected the defendant expert's evidence, substantially on the basis that the claimant denied any such attacks and there was no evidence to undermine his contentions on this point. More importantly a link was established on each occasion between the presence of a higher than normal level of chemical fumes and the claimant's symptoms. Specifically the judge noted that the presence of

28. [1956] AC 613.

29. The same point was made in *Nicholson* v *Atlas Steel Foundry and Engineering Company* [1957] 1 WLR 613.

30. [2000] Lloyds Rep Med 254; [2000] 3 All ER 421.

wheezing on each occasion when the claimant collapsed was consistent with asthma. Causation was therefore made out.

Expert Evidence

Expert evidence will be required on breach of duty and causation. Whilst an expert engineer will have some knowledge of the industrial processes that use or create hazardous substances, it is important to ensure there is a particular expertise in the context of industrial asthma. The expert needs to have a thorough knowledge of the industrial or chemical process in question. In some cases an occupational hygienist may be the more appropriate expert.

The report, which has to be based on a site visit, needs to address the following matters.

(1) The knowledge of the risks of injury through exposure to the hazardous substances in question.

(2) An identification of the hazardous substance of which complaint is made.

(3) An analysis of the employer's process with an opinion on whether and to what extent the claimant came into contact with the hazardous substance and the nature and extent of that exposure.

(4) Having assessed the system of work it must state the measures taken to prevent exposure and whether these comply with the relevant statutory regulations, (e.g. a measurement of changes of air in a premises per 24 hours, the provision of protective equipment etc). If there is a breach of duty, statutory or at common law, then these have to be set out in some detail.

(5) If relevant, it has to assess what else could account for the claimant's symptoms.

(6) Causation: whether the alleged breach was causative of some injury and the reasons for that opinion.

Industrial Dermatitis: Overview of the Statutory Regulations

Many of the principles discussed in relation to industrial asthma will also apply to industrial dermatitis. There is a similar duty both at common law and pursuant to statutory regulation to prevent the exposure of the employee to substances hazardous to health. Statutory regulations will be relevant to industrial dermatitis, but any gaps in the duty of care are filled by the common law.

The principal statutory regulations to consider include:

(1) The COSHH Regulations 2002[31]

(2) The Workplace (Health, Safety and Welfare) Regulations 1992[32]

(3) The Provision and Use of Work Equipment Regulations 1992 and 1998

(4) The Personal Protective Equipment at Work Regulations 1992 and 1999.

Breach of Duty and Causation

As to the duty of care and the breach of that duty, the principles are illustrated in *Pape* v *Cumbria County Council*.[33] The claimant was a cleaner who used detergents and other cleaning materials in the course of his employment over a sustained period of time. He was supplied with rubber gloves but did not wear them. He contracted dermatitis on his hands that spread over his body to the extent that he could no longer continue with his job. It was held that the employer should have known of the risks and there was a negligent omission to warn the claimant of the dangers from these substances. The fact that gloves were provided did not of itself absolve the defendant of liability and it should have ensured the gloves were worn.

31. Or those regulations in force at the relevant time.
32. Particularly note the requirement for suitable washing facilities: Reg. 21.
33. [1992] All ER 211.

As to causation, the same principles will apply as set out in relation to asthma. *McGhee* v *NCB*[34] was a case relating to industrial dermatitis, where the allegation was based on a lack of proper washing facilities in the context of a dusty industrial environment. Proof of causation was complex in that it could not be said that the breach of duty of the defendant had caused the dermatitis, but rather it had significantly increased the risk of injury. This increase in risk was held to equate to a "material contribution" to the injury that entitled the claimant to claim for the entire extent of his loss.[35] For the reasons set out in Chapter 2 on causation, this analysis may no longer hold good in this type of case. Instead a claimant has to prove that the breach of duty has made a material contribution to the injury and a material increase in the risk of injury by the breach of duty will not be sufficient.[36] Thereafter the defendant's contribution to the injury must be assessed under the principles outlined above.

Quantum of Damages for Dermatitis and Occupational Asthma

Set out on page 248 is a series of cases that provides some guidance on the awards that are made for these injuries. Practitioners should also consider the following issues in assessing damages:

(1) Are provisional damages relevant in the case of a "risk of serious deterioration" in the future condition? If the condition is progressive such that a medical opinion can provide an accurate prognosis at trial then the award will probably not be made.

(2) An assessment of constitutional and other causal factors (e.g. smoking) that could or would have affected the claimant's condition in any event.

(3) Additional "bolt on" expenses/losses arising from the conditions (e.g. creams in the case of dermatitis) on a lifetime basis.

(4) Whether there is a handicap in the labour market.

34. [*Supra*].
35. Namely that " a material increase in the risk of injury" was considered equivalent to " a material contribution to the injury" thereby satisfying causation.
36. See, generally, Chapter 2 on causation.

(5) Medical evidence has to cover these points in detail.

Awards for Industrial Asthma and Dermatitis

Name of Case	Facts and Award
Goodenough v *Dunlop Ltd* [1996] 96[3] Q.R. 7	Male, 47 employed as a process worker. Suffered asthma at work between 1963–1988. He was exposed to fumes of Tolylene di-isocyanate, ("TDI"). Held 80% of the claimant's asthma was occupational with 20% being due to constitutional factors. G's asthma put him at risk of an ongoing daily disability. His life expectancy was reduced by approximately five years. Condition was chronic and improvement was unlikely. Disablement assessed at 30%. *General Damages: £5,000. Damages for disadvantage on the labour market: £25,000.*
Thomas v *West Glamorgan* AHA [1996] Current Law 2298	Female, 34 employed as a domestic assistant at a hospital where she came into contact with detergents. Provided with rubber gloves, a long sleeved overall and an apron for protection, but the detergent ran down the gloves and into contact with her skin. She was diagnosed with dermatitis. She would suffer from hand dermatitis indefinitely and extra care was required in protecting her hands. *General Damages: £6,000. In addition: £50pa for the purchase of gloves and handcream.*

Name of Case	Facts and Award
Sola v *Royal Marsden Hospital* [1997] Current Law 1948	Female 51 was exposed to glutaraldehyde ("Cidex") at work as a nurse. Suffered breathlessness and lethargy. Occupational asthma was diagnosed. She was moved to a different employment but her exposure continued. She became depressed and was diagnosed with "compassion fatigue". She was retired on medical grounds. She only suffered asthmatic symptoms if she had contact with glutaraldehyde, but still carried an inhaler. *General Damages: £14,000. Award for loss of congenial employment: £5,000.*
Pape v *Cumbria County Council* [1992] All ER 211	Female (57) employed as a cleaner using detergents. Did not wear rubber gloves provided and contracted dermatitis. Spread over her body. Had to give up work. Her weight increased and she had to take steroids. Resultant erythroderma: *General Damages: £22,000.*
Fashade v *North Middlesex Hospital NHS Trust* [2000] Wright J: QBD Unreported[37]	Female nurse was exposed to glutaraldehyde when sterilising equipment. She suffered coughing fits, nose and eye irritation. She developed respiratory difficulties, including breathlessness and she was retired on medical grounds. Assessed at 30–40% disabled: 7.5% risk of 80% respiratory failure in her mid seventies. No provisional damages awarded. *General Damages: £30,000.*

37. Reported in JPIL, June 2001.

Summary of this Chapter

To provide practical guidance on the handling of these claims the schedule below highlights the relevant points from both a claimant and defendant practitioners' perspective.

Issues to be covered or noted	Claimant's Perspective	Defendant's Perspective
Duty of care Foreseeability. The date of knowledge of the risk of injury is important to assess the extent of the duty of care.	Identify the substances in issue and prove the defendant should have known of the risks of injury. Expert evidence is required.	Is knowledge of the risk of injury relevant? What is the evidence of risk on the identified substances. Cover the point on the expert evidence. Clarify the situation with the manufacturers.
Breach of statutory duty: COSHH 2002 regulations. Also the "six pack 1992" regulations are relevant. • Employers have to provide proper protective equipment and ensure it is worn. A known significant risk of injury possibly warrants removal of the employee.	• Disclosure of all relevant documentation on risk assessments under COSHH. • Assess all other documents on system of work and precautions required.	• Establish all risk assessments and compliance with COSHH. • Consider all documentation on the system of work. What precautions were required and what precautions were taken? • Was there an avoidable exposure? Dismissal of the employee required?
Limitation: Claims can be time barred and this must be considered.	• Is the claim time barred? • How long has the claimant been aware of the injury? • Assess the medical records.	• Investigate all medical records, GP notes and DSS/Benefits Agency medical examinations.

Issues to be covered or noted	Claimant's Perspective	Defendant's Perspective
Evidence of Breach of duty: An expert who is properly qualified must report on breach of duty and causation.	• Ensure the expert is properly qualified. Is there a particular expertise required? (e.g. chemical exposure).	• Ensure the expert is properly qualified and knows the claimant's case. • Ensure that senior staff can also provide supportive evidence so "gaps" in the evidence are avoided. • Note "reasonable practicability" qualification under the new regulations.
Evidence: Essential requirements for a diagnosis of occupational disease are: (1) Exposure to a sensitising agent. (2) Whether the condition improved when away from work. (3) Whether there had been prompt recurrence on re exposure. (4) The time lag between the start of exposure and the onset of symptoms.	• Expert evidence needs to be based on a statement from the claimant. The precise nature of any breach. What precautions should have been taken? • Was this breach likely to cause some injury?	• Expert evidence is required on these points. • Were adequate precautions taken? • What is the evidence of other colleagues? • What is the history of complaints? • Was the breach likely to cause injury or was it de minimis?

Issues to be covered or noted	Claimant's Perspective	Defendant's Perspective
Causation: Claimant has to prove the exposure caused some injury. A "material contribution" to the injury is required. Medical evidence is needed. As to the law: see *Holtby* and *Fairchild*.	• Ensure the expert evidence establishes medical causation. Are there constitutional factors that compromise causation? • The nature and extent of such factors has to be considered on the medical evidence.	• Expert evidence is essential. Can the symptoms be explained on any other basis? • Is it asthma/dermatitis and is the system of work causing the symptoms? • Are there pre-existing factors that explain the symptoms?
Damages: Assess contribution to the injury of all causes. Is this a divisible injury? • Assess any future risk of deterioration. Provisional damages and other future losses can be relevant.	• Clarify the condition and prognosis, including the long term nature of the condition. • Will the claimant have to give up work or suffer a deterioration? • Assess the percentage chances and consider provisional damages.	• Can the claimant continue work and what is the percentage chance of a deterioration? • Avoid provisional damages if possible, potentially on the basis that the condition is progressive. • Is a future loss claim too "speculative"? Medical evidence may assist.

SECTION 2: THE MEDICAL PERSPECTIVE

Overview

In 1989 a national surveillance scheme for work-related respiratory diseases was set up in the UK: the Surveillance of Work Related Occupational Respiratory Diseases ("SWORD"). This soon demonstrated that occupational asthma was the most common type of occupational respiratory disease. Workers involved in coach and spray painting produced the largest number of cases.

Basic Pathophysiology

The asthmatic response

Asthma is defined as a disease characterized by increased responsiveness of the bronchi to various stimuli, which manifests itself as widespread narrowing and obstruction of the airways, which is reversible, either spontaneously or with drug treatment. This is in marked contrast to the irreversible airway obstruction of conditions such as chronic bronchitis and emphysema.

There are three components:

(1) The acute asthma attack itself;

(2) The abnormally heightened response of the airways ("hyper-reactivity") to a number of stimuli, such as cold air, exercise, upper respiratory tract infections and other irritants;

(3) Persistent inflammation of the airways.

Bronchial hyper-reactivity

The bronchial hyper-reactivity lies at the heart of the asthmatic response. It is a unique property of asthmatic airways to react by bronchoconstriction, inflammation and mucus production to stimuli such as cold air, that would not elicit such a response in normal airways.

Immune mechanisms

Immunological mechanisms play an important part in this highly complex process. An immediate response to proteins or chemicals that are perceived to be "foreign" is mediated by cells lining the airways known as "mast cells".

These cells liberate histamine and a number of other chemical factors that result in an immediate inflammatory reaction. This results in the typical symptoms of acute asthma, including cough, wheezing and breathlessness.

Other chemical mediators liberated by white cells known as lymphocytes, such as "leukotrienes" and "platelet activating factor", result in a delayed immune response, which develops around four to six hours after the immediate reaction. Neural mechanisms, involving the involuntary or "autonomic" nervous system, also play an important role in maintaining the calibre of the airways and their secretions.

It is these chemical and neural mechanisms which result in prolongation of asthmatic symptoms after exposure to the "foreign" protein or agent has ceased.

Irritant-induced Occupational Asthma

The response of the lungs to irritants

Agents inhaled at work may initiate chronic asthma by causing toxic damage to the delicate cells lining the bronchial airways (mucociliary epithelium) after inhalation of a respiratory irritant in toxic concentration (irritant induced asthma). Examples of irritants which may initiate asthma include:

(a) Chlorine

(b) Ammonia

(c) Sulphur dioxide

(d) Oxides of nitrogen.

Agents which *initiate* asthma need to be distinguished from those which *provoke* asthmatic symptoms in individuals with pre-existing hyper-reactive airways, for example sub-toxic levels of sulphur dioxide or smoke.

Clinical features

In the case of irritant-induced asthma, there is no history of pre-existing respiratory disease. The symptoms of asthma usually develop within 24 hours of a single specific incident involving inhalation of irritant gases, fumes, vapours or smoke in very high concentrations.

The common symptoms include cough, wheezing and shortness of breath. The symptoms will usually have persisted for at least three months. Examination of the chest may confirm the presence of wheezing, which tends to be more prominent during expiration.

Investigations

Investigations which are commonly undertaken include:

(a) *Formal pulmonary function tests*, such as the volume of air forcibly expired after a full inspiration (FVC), and the volume of air that can be forcibly expired in one second (FEV_1). The typical obstructive pattern seen in asthma is for both the FVC and the FEV_1 to be reduced, but with a disproportionate reduction in FEV_1;

(b) *Bedside pulmonary function tests*, such as the peak expiratory flow rate (PEFR), which can be measured using a simple hand-held peak flow meter;

(c) *Chest x-ray*. Although often normal in asthma, sometimes there is evidence of hyperinflation, and characteristic changes caused by plugs of mucus;

(d) *Inhalational challenge testing* – measuring pulmonary function after exposure to the suspected irritant in controlled conditions.

Hypersensitivity-induced Occupational Asthma

The role of sensitisers

Occupational asthma may also occur as a result of an acquired allergic reaction to a sensitizing agent (a "sensitiser"). This is known as hypersensitivity-induced asthma.[38] Examples of important inhaled sensitisers in the occupational setting include:[39]

(a) Diisocyanates[40]

(b) Flour

(c) Grain dust

(d) Latex

(e) Rat and mouse urine proteins[41]

(f) Soldering flux (colophony)

(g) Enzymes[42]

(h) Acid anhydrides[43]

(i) Complex platinum salts.

Clinical features

The clinical history usually shows a characteristic pattern:

(a) An initial symptom-free period;

(b) Weeks or months, but sometimes years after exposure to a known sensitiser, symptoms of asthma occur during the working week, and may worsen as the week progresses;

38. There is often an associated allergic rhinitis, producing chronic nasal discharge.
39. For a fuller list, see the Prescribed Industrial Diseases, at PD D7.
40. Used in the manufacture of polyurethanes.
41. The usual source is laboratory animals.
42. Used in detergents and the food industry.
43. Used in the manufacture of epoxy resins.

(c) The typical symptoms of cough, wheezing and shortness of breath improve at weekends, when on holiday or when absent from work for other reasons.

Investigations

The most important investigations include:

(a) Serial PEFR measurements, based upon the best of three measurements made every two hours from waking to sleeping over a one month period, including a period longer than a weekend away from work. A significant drop in PEFR or substantial variability on working days, but not on days away from work, supports the diagnosis;

(b) Immunological investigations, including skin testing with extracts of the sensitisers which commonly cause occupational asthma, and assay of specific IgE;[44]

(c) Inhalational challenge testing[45] tends to be reserved only for cases where there is diagnostic doubt after the above investigations.

Management of Occupational Asthma

Avoidance

The mainstay of management is avoidance of further exposure to the sensitiser. If exposure cannot be completely prevented, it should be minimized, and respiratory protection worn. However, if asthmatic symptoms recur in spite of these measures, it may be necessary to provide the affected worker with alternative work, where this is possible.

Medical therapy

The conventional management of asthma includes the use of inhaled bronchodilator drugs such as salbutamol, inhaled steroids, and short courses of oral steroids when symptoms acutely deteriorate.

44. An antibody protein to the sensitiser found in the serum. It is specific to the suspected substance.

45. Reproduction of symptoms in controlled conditions, following inhalation of the suspected sensitiser.

Prevention

In some cases reduction to exposure requires only simple measures, such as adequate respiratory protection, ventilation and air filtration. The substitution of enzyme granules for powder is an example of another simple measure which resulted in a significant drop in airborne exposure.

However, in many other cases measures to reduce exposure have been inhibited by lack of knowledge of exposure-response relationships, and difficulties in measuring the concentrations of airborne substances, particularly proteins such as those found in rat urine. In more complex cases such as these, control measures may also need to be directed to factors such as cage design, bedding and humidity.

Occupational asthma following exposure to a prescribed agent is a prescribed industrial disease (PD D7). Allergic rhinitis due to a prescribed agent is also a prescribed industrial disease (PD D4).

Further Reading

Newman Taylor, A. "Occupational Asthma" in "Hunter's Diseases of Occupations" (9th edition). London, Edward Arnold, 2000.

Madan, A. "Occupational Asthma and Other Respiratory Diseases". In: *ABC of Work Related Disorders*. London, BMJ Publishing Group, 1997.

Occupational Dermatitis Overview

There are many known occupational skin diseases (the "dermatoses"), however, the large majority are dermatitic in nature, in other words they result in dermatitis. Most cases of occupational dermatitis involve the hands.

Contact Dermatitis

Contact dermatitis is dermatitis caused by skin contact with external agents. This form of dermatitis needs to be distinguished from constitutional (endogenous) dermatitis. It may arise due to skin contact with irritant substances (irritant contact dermatitis), or contact with substances which induce an allergic reaction (allergic contact dermatitis).

Irritant contact dermatitis

The outer layer of skin is known as the epidermis. The very outermost layer (the "stratum corneum") is formed of layers of dead cells, which become thickened at areas of friction and pressure, such as the palms of the hands, and the soles of the feet.

The epidermis normally provides an effective barrier against weak chemical and physical agents. However, if an irritant remains in contact with the skin for sufficient time or at sufficient concentration, the living cells of the skin will eventually become damaged.

The typical acute symptoms include redness, itching, and small blisters filled with clear fluid or pus. Over time, the skin will eventually become dry and thick, with painful fissures. In a mild form it is often referred to as "chapping". There is great individual variation in terms of susceptibility. Sometimes many years may elapse before symptoms occur.

The higher risk occupations include those which result in frequent "wet" hands, such as hairdressing, catering and nursing. Other higher risk occupations include agriculture, vehicle maintenance, construction, printing and metalworking.

Examples of irritants found in the workplace include:

(a) Soaps and detergents

(b) Acids and alkalis[46]

(c) Organic solvents

(d) Metalworking fluids[47]

46. For example, sodium hydroxide used in degreasing liquids.
47. For example, cutting oils.

(e) Animal and plant products[48]

(f) Oxidising agents.[49]

Allergic contact dermatitis

If a contact sensitiser penetrates the barrier layer of the skin, it may provoke a delayed allergic reaction. After exposure, there is a latent or "induction" phase during which hypersensitivity to the substance is acquired. This may follow one exposure, or multiple exposures over a prolonged period. The induction period usually lasts for about seven to ten days.

Once induction has occurred, re-exposure to the sensitiser results in acute inflammation, and the typical symptoms of dermatitis, usually about one to day days after re-exposure.

Many thousands of sensitisers have now been recognized. The agents commonly implicated include:

(a) Natural rubber latex[50]

(b) Chemicals used in rubber manufacture.

(c) Sterilising agents[51]

(d) Thioglycolates[52]

(e) Epoxy resins[53]

(f) Chromates[54]

(g) Plant products.

48. For example, vegetable juices.
49. For example, hydrogen peroxide, which is found in many products used to lighten hair.
50. As found in gloves used by health care workers.
51. For example, formaldehyde (formalin).
52. Commonly found in hairdressing chemicals.
53. Used in the manufacture of many plastics and glues.
54. Found in many cements.

Diagnosis

Establishing the cause of occupational contact dermatitis is not always straightforward. Relevant factors include:

(a) *Site:* involvement of the hands is typical. However, the same pattern may occur with endogenous forms of dermatitis.

(b) *Time of onset*: whether the dermatitis first occurred in an occupational setting.

(c) *Occupation*: whether the occupation is one that results in exposure to known irritants or allergic sensitisers.

(d) In the early stages, an improvement or complete clearance of symptoms when *away from work*. Allergic contact dermatitis tends to relapse more rapidly upon to return to work than irritant contact dermatitis.

(e) *Responses of colleagues*; the involvement of a large number of other employees tends to suggest irritant contact dermatitis.

(f) *Patch testing of skin* with known sensitisers. This is of no use in cases of irritant contact dermatitis, but may confirm the diagnosis in an allergic case.

(g) Sometimes *specific blood tests and biopsy* of the affected area of skin are required.

Treatment

The mainstay of treatment is complete or partial removal from the irritant or sensitiser, together with skin preparations such as steroid ointments to reduce inflammation, and emollients (moisturizers), to reduce dryness, cracking, and provide a waterproof barrier.

In general, cases of irritant contact dermatitis respond quite well to emollients, whereas allergic contact dermatitis will not usually resolve without steroid ointments as well. In some resistant cases, where exposure to irritants or sensitisers cannot be avoided, a change of career may become necessary.

Prevention

Pre-employment health screening may be of use in identifying individuals who are at higher risk, for example those with a history of severe childhood eczema, who are probably more susceptible to skin irritation.

Personal protective equipment, in particular gloves, may have a role to play. However, in some occupations gloves may be neither safe nor practicable.

The role of "barrier" creams has been controversial, however, they may reduce the exposure to skin contact, particularly from irritants. Emollients applied after skin cleaning may also have a role to play.

Once symptoms have developed, it is sometimes possible to substitute the offending irritant or sensitiser with another agent, which lacks the same potential to cause dermatitis. In more stubborn cases, a change of occupation may be required. In a small minority of patients, significant symptoms persist long after cessation of exposure.

Occupational dermatitis is a prescribed industrial disease when caused by exposure to dust, liquid, vapour or any other external agent (PD D5).[55]

Further Reading

Rycroft, R. "Occupational Diseases of the Skin" in *Hunter's Diseases of Occupations* (9th edition). London, Edward Arnold, 2000.

White, I. "Occupational Dermatitis". In: *ABC of Work Related Disorders*. London, BMJ Publishing Group, 1997.

55. Except for dermatitis caused by exposure to chromic acid, chromates or bi-chromates (PD C30).

CHAPTER 12
INDUSTRIAL DISEASES AND THE CORONER'S INQUEST

Reporting deaths to the coroner

Certain individuals have a duty imposed on them to report deaths to the coroner or their officer.[1] In other cases, it is the duty of those persons who are "about the deceased" to report circumstances which may require the holding of an inquest to a coroner or their officer immediately. A death due to a suspected industrial disease would amount to circumstances which may require the holding of an inquest.

If the deceased was a member of a trade union at the date of death and was still employed, there is also a duty to report to the enforcing authority (usually the HSE) a death caused by a prescribed industrial disease under the Social Security (Industrial Injuries) (Prescribed Diseases) Regulations 1985, SI1985/967.

The need for a post mortem examination

Where he or she is informed that a body is lying within their district a coroner may take the following action:

(a) Decide that a post-mortem examination or inquest is unnecessary and notify the Registrar of Deaths;[2]

(b) Where there is *reasonable cause* to suspect that the person has died a *sudden death of unknown cause*, direct a post-mortem examination, and on receipt of the report decide

1. For example the governor or manager in the case of deaths in prison, in detention centres or in a young offenders' institute.
2. Using "Pink Form A".

that an inquest will be unnecessary and notify the Registrar of Deaths.[34]

Where a coroner directs a post mortem examination, he or she is obliged to notify the following persons or bodies of the time and place of the examination, unless impracticable or to do so would cause undue delay:[5]

(a) Any relative of the deceased who has notified the coroner of their desire to attend or be represented[6]

(b) The deceased's regular medical attendant[7]

(c) If the deceased died in hospital, the hospital[8]

(d) If the death of the deceased may have been caused by a prescribed industrial disease, the medical boarding centre (respiratory diseases) for the area[9]

(e) If the death of the deceased may have been caused by any accident or disease, notice of which is required under any act, the enforcing authority or inspector appointed[10]

(f) Any government department which has notified the coroner of its desire to be represented[11]

(g) The chief officer of police if he has notified the coroner of his desire to be represented.[12]

In the case of deaths which may be due to a prescribed industrial disease, the coroner should not direct that the post mortem examination be carried out by a pathologist who is a member of a medical boarding centre (respiratory diseases); CR r 6(1)(d). However, the Home Office has advised[13] that:

(a) The pathologist should be sufficiently experienced, and have facilities for "fixing" lung tissue.

3. Using "Pink Form B".
4. Coroners Act 1988 ("CA") s19(3).
5. Coroners Rules 1984 ("CR") r7(1).
6. CR r7(2)(a).
7. CR r7(2)(b).
8. CR r7(2)(c).
9. CR r7(2)(d).
10. CR r7(2)(e).
11. CR r7(2)(f).
12. CR r7(2)(g).
13. Circular No. 18 of 1980.

(b) Coroners should co-operate with medical boarding centres (respiratory diseases), and take advantage of their expertise, by notifying them of such deaths and supplying them with a copy of the post mortem examination report, when available, together with access to the relevant organs and other useful pathological material, or authorising the pathologist to do so.

(c) Where possible, differing conclusions should not be reached by coroners (and their pathologists) on the one hand, and medical boarding centres (respiratory diseases) on the other.[14]

The need for an inquest

An inquest *must* be held where the coroner has *reasonable cause* to suspect:

(a) Violent or unnatural death[15]

(b) Sudden death of cause unknown[16]

(c) Death in prison or such other place[17] or in such circumstances to require an inquest under any other Act.

"Unnatural" is given its ordinary meaning and whether a death is due to natural or unnatural causes is a matter of fact and common sense in each case.[18] There is no discretion as to whether to hold an inquest or not if the case clearly falls within s 8(1) CA 1988. Thus, if the coroner has reasonable cause to suspect that death was due to an industrial disease, then this will be "unnatural", and an inquest must be held.[19]

Conversely, there is no duty, and therefore no power, to hold an inquest in the case of deaths which the coroner has reasonable

14. For an example of a mesothelioma case where different conclusions were reached see: *Terry* v *HM Coroner for E Sussex* [2001] 3 WLR 605 (CA).
15. CA s8(1)(a).
16. CA s8(1)(b).
17. CA s8(1)(c); even where the death was apparently due to natural causes.
18. See *R* v *Poplar Coroner ex p Thomas* [1993] QB 610.
19. Where a coroner has reasonable cause to suspect that the deceased's death was at least contributed to by "neglect" (see below) then it will be "unnatural", and an inquest must be held. Similarly, whenever a coroner has reasonable cause to suspect that a wholly unexpected death, albeit from natural causes, resulted from some culpable human failure, then an inquest probably ought to be held. See *R* v *HM Coroner's Court for St Pancras ex p Touche* [2001] 3 WLR 148 (CA).

cause to believe are due to *natural causes*.[20] In practice, this will often not be known until the coroner is in receipt of the post mortem.

The purpose of the inquest

An inquest attempts to establish:

(a) Who the deceased was

(b) How,[21] when and where the deceased came by his or her death

(c) The particulars required by the Births and Deaths Registration Act 1953 to be registered.

Note that this is an inquisitorial fact-finding exercise, and not an adversarial process of apportioning guilt or fault. In theory at least, there are no parties or sides. Furthermore, a verdict must not be framed so as to appear to determine any individual criminal or civil liability.[22]

Duty to summon jury

A coroner *must* summon a jury if he or she has reason to suspect that:

(a) Death occurred in prison or in such a place or in such circumstances as to require an inquest under any other Act[23]

(b) Death occurred while the deceased was in police custody, or resulted from an injury caused by a police officer in the purported execution of his or her duty[24]

(c) Death was caused by an accident, poisoning or disease notice of which is required to be given under any Act[25]

20. See CA s19(4).
21. See CA s11(5)(b)(ii) and CR r36(1)(b).
22. CR r42. See also *Jamieson* and *Middleton* [infra].
23. CA s8(3)(a).
24. CA s8(3)(b).
25. CA s8(3)(c). See also the Reporting of Injuries, Diseases and Dangerous Occurrences Regulations 1995 SI 1995/3163; the Social Security (Industrial Injuries) (Prescribed Diseases) Regulations 1985, SI1985/967.

(d) Death occurred in circumstances, the continuance or possible recurrence of which is prejudicial to the health or safety of the public.[26]

A coroner also has a discretion as to whether to summon a jury where it appears to him or her that there is any other reason for doing so.[27] It is important to note that the reporting criteria under (c), hence the need for a jury inquest, do not apply where the deceased had retired, as will often be the case in diseases such as mesothelioma, which may arise many years after the relevant employment has ceased. Similarly, these provisions will not apply if the deceased was still working, but did not meet the strict occupational criteria.

Witnesses and evidence

An inquest is normally opened within a few days of death, in order to deal solely with identification. The inquest is then adjourned to allow time for further investigations.

(1) Witnesses

A coroner has a discretion to decide which witnesses to call to give evidence. He or she will examine on oath those persons tendering evidence as to the facts of the death and all persons having knowledge of those facts, who he or she considers it "expedient" to examine.[28] However, a coroner also has to be fair and consider all applications for witnesses to be called in the light of the matters to be determined at an inquest.

Note that a witness at an inquest does not have to answer any questions which may tend to incriminate him or herself.[29] Further, there is no "rule" that prevents a coroner from admitting hearsay evidence, in either written or oral form.

(2) Documentary evidence

A coroner has a discretion to admit documents, which he or she considers to be relevant and properly admissible, unless an interested person objects. If objections are raised, the document

26. CA s8(3)(d). The onus appears to be on the applicant to show that the death occurred in a systemic context, as opposed to a one-off fatality.
27. CA s8(4).
28. CA s11(2).
29. CR r22(1).

may still be admitted, if the coroner determines that that the maker is unable to give oral evidence within a reasonable period.[30] Any documentary evidence which a coroner admits at an inquest must be read aloud, unless the coroner directs otherwise.

In practice, only non-contentious documents, including uncontroversial witness statements, will usually be read. On any issue which requires more careful consideration, it is expected that oral evidence will be given by the person who made the note.[31] Any properly interested person is also entitled to receive copies of any documents put in evidence at an inquest.[32]

(3) Problems with advance disclosure

A coroner has no power to order the production of documents, although as a last resort he or she may apply to the High Court for a *subpoena duces tecum* ordering their production at the inquest.

In practice that is not usually necessary, and bodies such as the Chief Constables, the Police Complaints Authority, the Prison Service, the Health & Safety Executive, and the Crown Prosecution Service will usually furnish the coroner with statements, investigation reports and other relevant documents on a voluntary basis.

However, the coroner has *no power to disclose such documents to a third party,* and limited powers to prevent such disclosure. Similarly, an interested party has no right to call for the advance production of documents, including witness statements, which have been disclosed to the coroner in confidence.[33] It is for the body commissioning the report and/or the maker of the statement to decide whether to do so voluntarily, and if so, on what terms.

Nevertheless it is improper for a coroner to use any material which is not made available to interested parties, primarily the family of the deceased.[34] On the ground of fairness, a party

30. CR r37. In practice, this means that the document must be disclosed in advance, even if at the hearing.
31. *R* v *Southwark Coroner ex p Hicks* [1987] 1 WLR 1624.
32. CR r57.
33. *R* v *HM Coroner sitting at Hammersmith ex p Peach* [1980] QB 211 (CA). Significant criticism of the principle, which is out of step with contemporary practice in the civil courts under the CPR, and which frustrates the common law requirement for fairness to the parties, was voiced by Sullivan J in *R* v *HM Coroner for District of Avon ex p Bentley* [2001] EWCA Admin 170.
34. *R* v *HM Coroner for Ceredigion ex p Wigley* [1993] COD 364.

which may be held responsible for the death should also see such documents.[35]

The modern trend is for the coroner to release relevant copy documents or statements upon request prior to the hearing where at all possible.[36] Certain documents can normally be obtained in advance in any event, for example:

(a) The post-mortem examination report[37]

(b) Medical records

In the light of the Lawrence Inquiry, detailed protocols now exist, which provide guidance on advance disclosure to both the Prison Service and Chief Officers of Police in the case of deaths in prison and deaths in custody.[38] There are no such protocols in force with regard to deaths at work or due to an industrial disease. However, where they have become involved, some HSE inspectors or investigating police officers are prepared to discuss their findings informally in advance of the hearing.

Persons entitled to examine witnesses

A person wishing to examine a witness ("interested persons") must satisfy the coroner that they are one of the persons listed in rule 20(2) of the Coroners Rules 1984.[39] These include:

(a) A parent, child, spouse or personal representative of the deceased[40]

(b) Any person whose act or omission may have caused or contributed to the death

35. See *Hicks* [*supra*].
36. The Coroners' Society has published guidance on advance disclosure in the "Practice Notes for Coroners" (19.11.98) at para 7. Where the deceased has died in police or prison custody or whilst detained under the Mental Health Act, they advise that any application should be considered very carefully and acceded to if at all possible.
37. CR r57.
38. See for example, Home Office Circular 31/2002.
39. See *R* v *Greater London South District Coroner ex p Driscoll* (1995) 159 JP 45 and *R* v *Coroner of the Queen's Household ex parte Al-Fayed* (2001) 58 BMLR 205 for further guidance. "Properly interested" denotes that the interest must be reasonable and substantial, not trivial or contrived.
40. This does not include the deceased's brothers or sisters, and is limited to spouses rather than other forms of partnership. However, they will normally be deemed "interested", under the "any other person" category.

(c) Any person appointed by the deceased's trade union if death was due to an injury or illness received in the course of employment

(d) An inspector appointed by or representative of an enforcing authority or government department

(e) Any other person who in the opinion of the coroner is "properly interested."

They can examine witnesses either in person or by counsel/solicitor. Interested parties or their representatives may also make submissions of law and submissions as to an appropriate verdict. However, they are not allowed to address the coroner or the jury as to the facts.[41]

Adjournments

The coroner must adjourn an inquest:

(a) Where the death was due to industrial accident or disease and an inspector or representative of an enforcing authority, such as the Health & Safety Executive, is not present[42]

(b) Where a person's conduct may be called into question and he or she has not been summoned to attend or given notice of the inquest[43]

(c) Where there are pending relevant criminal charges and the chief officer of police[44] or the DPP[45] request an adjournment.

The coroner will also adjourn during an inquest of his or her own motion if the death is likely to be due to any of the criminal offences set out in CR r26(3).[46]

41. CR r40.
42. CR r23.
43. CR r25.
44. CR r26(3).
45. CR r27.
46. CR r28.

The scope of the inquiry

The scope of the inquiry, "how" the deceased came about his or death, is a controversial area. In addition to the discretion to restrict the witnesses to those "which it is expedient to examine", the coroner has a discretion to disallow any question which in his or her opinion is not relevant or is otherwise not a proper question.[47]

The manner in which the coroner exercises these discretionary powers will depend on what the word "how" means to them. However, notwithstanding the limitations imposed by rules 20(1)(b), 36(1)(b) and 42, the coroner has to ensure that the relevant facts are fully, fairly and fearlessly investigated. All aspects touching upon negligence or unlawful killing may therefore be investigated.[48]

Verdicts and further action

There is no statutory or common law provision that a verdict has to be returned in a particular form.[49] Some coroners prefer to use "narrative verdicts", brief, neutral, factual statements. For example: "The deceased drowned when his yacht capsized in heavy seas". The short form verdicts listed in the notes to Inquisition Form 22 are merely suggestions. Examples include:

(a) Death from natural causes

(b) Death as a result of accident/misadventure[50]

(c) Death from industrial disease (disease specified)

(d) Open verdict.

47. CR r20(1)(b).
48. *R* v *HM Coroner for North Humberside and Scunthorpe ex p Jamieson* [1995] 1 QB 1.
49. *R* v *Southwark Coroner ex p Kendall* [1988] 1 WLR 1186.
50. There is no distinction in law; see *R* v *Portsmouth Coroner's Court ex p Anderson* [1987] 1 WLR 1640.

Until recently, *Jamieson*[51] was authority that self-neglect and neglect[52] could rarely, if ever, be appropriate verdicts on their own, that is to say "free-standing verdicts". In appropriate cases the verdict could include the *rider* "aggravated by neglect" or "aggravated by self-neglect", but only where there was a "clear and direct" causal connection between the conduct so described and the death. Those guidelines now need to be read in the light of the decision of the Court of Appeal in *Middleton*:[53]

After the evidence has been given, the coroner is obliged to hear legal submissions from interested parties or their representatives. If there is a jury, these will be heard without the jury. No person may address the coroner or the jury as to the facts,[54] however, when addressing a coroner as to what verdicts are appropriate, it is sometimes difficult to do so without some passing reference to the facts.

In jury cases, the coroner need not leave all possible verdicts to a jury, simply because there is technically evidence to support them. The coroner has a broader discretion than mere considerations of the strength of the evidence. It is sufficient that he or she leaves verdicts which *"realistically reflect the thrust of the evidence as a whole"*, since to leave all possible verdicts may confuse or overburden a jury.[55]

---

51. [*supra*].
52. A gross failure to provide adequate nourishment or liquid, or provide or procure basic medical attention or warmth or shelter to a dependent person (as a result of youth, age, illness or infirmity) who cannot provide it for him or herself, where a "clear and direct" causal connection is established between the conduct so described and the death.
53. *R* v *Secretary of State for the Home Department ex parte Amin, R* v *HM Coroner for West Somersetshire ex p Middleton* [2002] 3 WLR 505. In cases which involve breaches of article 2 of the Convention (the "right to life"), there is an obligation on the part of the State to conduct an "effective official investigation". Where the coroner's inquest is fulfilling that role, rule 42 could and should, contrary to *Jamieson*, when necessary be construed only as preventing an individual being named, with the result that a finding of free-standing "system neglect" would not contravene rule 42. The requirements are fact-specific, and will only apply where in the judgment of the coroner a finding of neglect could serve to reduce the risk of repetition of the circumstances giving rise to the death being inquired into. The coroner would in effect need to direct the jury when they could return a rider identifying the nature of the neglect they had found.
54. CR r40.
55. *R* v *Inner South London Coroner ex parte Douglas-Williams* [1999] 1 All ER 344, in which Woolf MR endorsed but qualified the test in *R* v *Galbraith* [1981] 1 WLR 1039.

Notwithstanding the constraints in relation to determining criminal or civil liability,[56] where a coroner believes that action should be taken to prevent a similar fatality, he or she may also announce at the inquest that they are reporting the matter in writing to the person or authority with the power to take such action.[57]

In the case of deaths involving prescribed industrial diseases, the verdict will usually be either natural causes or death from a specified industrial disease. Considerations of "system neglect" in the *Middleton* sense are unlikely to arise. Neglect riders in the *Jamieson* sense are also unlikely to arise, unless gross failures of medical care contributed to death.

Summary

The important aspects of this chapter with regard to industrial diseases are summarised below:

(a) Where a coroner has reasonable cause to suspect that the deceased may have died of an industrial disease, the death will be "unnatural", and an inquest must be held.

(b) If the deceased was a member of a trade union and still employed at the date of his death, there is a duty to report prescribed industrial diseases to the relevant enforcing authority.

(c) The coroner must also notify the date and place of the post mortem examination to the enforcing authority and the relevant medical boarding centre (respiratory diseases), and their right to be represented.

(d) The coroner may not instruct a pathologist who is a member of a medical boarding centre (respiratory diseases).

(e) The coroner is expected to make available the post mortem examination report, relevant organs and other useful pathological material to the medical boarding centre (respiratory diseases).

(f) The inquest must be held with a jury.

56. CR r42.
57. CR r43.

(g) The inquest must be adjourned if the inspector appointed by the enforcing authority is not present.

(h) The deceased's next of kin, the inspector appointed by the enforcing authority and the deceased's trade union are all "interested persons" for the purposes of examining witnesses in person or through their legal representatives.

(i) The coroner should avoid where possible reaching a conclusion which differs from that of the medical boarding centre (respiratory diseases).

(j) If the reporting criteria do not apply, for example where the deceased had retired, or, if still working, the strict occupational criteria were not met, then an inquest must still be held if the coroner has reasonable cause to suspect an unnatural death, but the notification criteria set out above are not required, and the inquest need not be held with a jury.

(k) The verdict will usually either be natural causes, or death by a specified industrial disease. A "neglect" rider will rarely arise in this context, unless there was a gross failure of the medical treatment afforded to the deceased, which is causally related to the death.

(l) The coroner has powers to make written recommendations to the appropriate body, where it appears that action needs to be taken to prevent a similar fatality.

APPENDICES

APPENDIX 1
MEDICAL SHORTHAND AND ACRONYMS

Note: many items are context dependent

A

A,a	attended GP surgery, blood group A, age, arm, artery
AA	Alcoholics Anonymous
A&0	alert and orientated
A&W	alive and well
AB	blood group AB
Abd, Abdo	abdomen, abdomen normal, abduction
ABG	arterial blood gases (a test of respiratory function)
abs	absent, abortions
Ac, a/c	Acute, acromio-clavicular, ante cibum (before food), air conduction (of sound)
ACE	acute coronary event
ACEI	angiotensin converting enzyme inhibitor (a class of drugs for high blood pressure)
ACTH	corticotropin
add, adn	additional, addiction, adduction
AE	air entry (of lungs)
A/E	Accident and Emergency Department
AF	atrial fibrillation (an abnormal heart rhythm)
AIDS	Acquired Immunodeficiency Syndrome
AJ, AJ+, AJ–	ankle jerk, ankle jerk present, ankle jerk absent
alc, alc+++	alcohol, intoxicated
AI	accidental injury
ALAMA	Association of Local Authority Medical Advisers
amps	ampoules
AN	antenatal (examination, clinic)
ANHOPS	Association of National Health Occupational Physicians
anx	anxiety
appt	appointment
ASD	atrial septal defect (a congenital heart lesion)
Assis	assistant

	assd	associated with
	audio	of the ear or hearing
	AV	anteverted (leaning forward)
B	B, b	bacillus, black, blood, blood group B, born, brother
	B6, B12 etc	B vitamins
	Ba, Ba m, Ba e	barium, barium meal/enema X-ray
	BA	Benefits Agency
	BAMS	Benefits Agency Medical Services
	BC	bone conduction (of sound)
	BCC	basal cell carcinoma (a skin cancer)
	BCG	bacille Camille Guérin (a vaccination against tuberculosis)
	BCS	battered child syndrome
	bd	bis die (twice daily)
	benzos	benzodiazepine drugs (eg Valium)
	BF	boy friend
	BID	brought in dead
	bilat	bilateral
	biochem	biochemistry estimations (eg of blood or urine)
	BKWP	below knee walking plaster
	blad	bladder
	bld	blood
	BMA	British Medical Association
	BMI	body mass index; weight kg ÷ (height m)^2
	BNF	British National Formulary (of drugs)
	BNO	bowels not open
	BO	bowels open
	BOA	born on arrival (in casualty)
	BP	blood pressure, British Pharmacopeia
	BPS	British Psychological Society
	BRI	Bradford/Bristol etc Royal Infirmary
	BS	bowel sounds, breath sounds, back slab plaster or splint
	Bx	biopsy
	BZ	benzodiazepine (eg Valium)
C	c, C	about, calorie, carbon, children, vitamin C, consultation, certificate (If underlined = with) C1..C2..C3 etc cervical vertebra or nerve, cranial nerve
	C_2H_50H	alcohol
	C&C	collar and cuff sling
	C & S	swab for culture and antibiotic sensitivity
	ca, Ca	carcinoma, cancer, calcium
	CABG	coronary artery bypass graft
	caps	capsules

CAS	Casualty Department
CCS	clean catch specimen of urine
CCU	Coronary Care Unit
CD	controlled drug
CDH	congenital dislocation of the hip
cert	certificate
chr	chronic
CIN	pre-malignant changes on a cervical smear
C/O	complains of
CO_2	carbon dioxide
Cons	Consultant, consultation, consciousness
CHD	coronary heart disease
CN	cranial nerves
CNS	central nervous system
COAD	chronic obstructive airways disease
COPD	chronic obstructive pulmonary disease
COSHH	Control of Substances Hazardous to Health
COT	College of Occupational Therapists
CPN	Community Psychiatric Nurse
CPR	cardiopulmonary resuscitation
creps	crepitations ie added crackles on auscultation of the chest
CRF	cryptogenic fibrosing alveolitis
C smear	cervical smear
CSF	cerebrospinal fluid
CSM	Committee on Safety of Medicines
CSP	Chartered Society of Physiotherapists
CSU	catheter specimen of urine
CT	chemotherapy
CTS	carpal tunnel syndrome
CT scan	computerised tomography scan
CVA	cerebrovascular accident (stroke)
CVP	central venous pressure
CVS, CVS_0	cardiovascular system, – normal
Cx	cervix
CXR	chest X-ray

D

D, d	date, daughter, day, degree, diagnosis, died, divorced, dry, duration, vitamin D
D1, D2, D3 etc	dorsal vertebra, dorsal nerve
D & V	diarrhoea and vomiting
Db	decibel (unit of sound)
DbA, dBA	A weighted decibels; weighted to the middle or more harmful frequencies
D&C	dilatation and curettage (of the lining of the womb)
DD	dry dressing
Dec'd	deceased

Dex	dextrose saline solution, dexamethasone
DGH	District General Hospital
DID	died of intercurrent disease
DIPF	diffuse interstitial pulmonary fibrosis
disloc	dislocated
displ	displaced
DL	danger list
DN	District Nurse
DNA	did not attend
DNR	do not resuscitate
DOA	dead on arrival
DOB	date of birth
DOD	date of death
DOEM	Department of Occupational & Environmental Medicine
DOH	Department of Health
DOI	died of injuries
DOHS	Department of Health Scotland
DOHW	Department of Health Wales
DS, D/S	dextrose saline
DSM	a system of classification of psychiatric disorders
DSS	Department of Social Security
DTP	diptheria, tetanus, pertussis (whooping cough) vaccine
DU	duodenal ulcer
DUI	driver under the influence
DV	domiciliary visit
DVT	deep vein thrombosis
D/W	discussed with
Dx	diagnosis
dysp	dyspnoea (shortness of breath)

E

E, e	ear, elbow, error, examination, eye
EAM	external auditory meatus (external ear)
ec	enteric coated (tablets)
ECG	electrocardiogram
ECT	electroconvulsive therapy
EDH	extradural haematoma
EEG	electroencephalogram
EMAS	Employment Medical Advisory Service
ENT	ear nose and throat
EOM	external ocular movements
EPI-DERM	a scheme of surveillance of occupational skin disease
epil	epilepsy
EPL	extensor pollicis longus tendon of thumb
EPB	extensor pollicis brevis tendon of thumb

ESN	educationally subnormal
ESR	erythrocyte sedimentation rate (a blood test of inflammation)
ERA	evoked response audiometry
ETA	estimated time of arrival
ETKM	every test known to man
EUA	examination under anaesthetic
EWO	ear/eye wash out
exp lap	exploratory laparotomy (opening of the abdominal cavity)
extn	extension

F

F, f	father, female, femur, fibula, finger, foot, fracture
FB, fb	foreign body (eg in the eye), fingers breadth wide
FBC	full blood count
FBS	fasting blood sugar
FDP	flexor digitorum profundis tendon of finger
FDS	flexor digitorum superficialis tendon of finger
Fe/FeSO4	iron tablets
FEV_1	forced expiratory volume of air in the first second
FFD	fixed flexion deformity
FH	family history
flex	flexion
FLKS	fatty liver and kidney syndrome
FNA	fine needle aspiration
FNF	fractured neck of femur
FMP	first menstrual period
FPA	Family Planning Association
FPL, fp	flexor pollicis longus tendon of thumb
FTA	failure to attend
FTLB	full term live birth
FIN	finger to nose test for brain function
FROM	full range of movement
FSH	follicle stimulating hormone
FTND	full term normal delivery
FTSG	full thickness skin graft
FTT	failure to thrive
FU	follow up
FUO	fever of unknown origin
FVC	forced vital capacity, the total volume of expired air
Fx	fracture
FYO	for your information

G

G, g	grade, gravida (number of pregnancies), gram
GA	general anaesthetic, Gamblers Anonymous
GGT	gamma glutaryl-transferase (a liver function test)
GB	gall bladder

	GCC	General Chiropractic Council
	GCS	Glasgow Coma Scale
	GDS	General Dental Council
	GH	growth hormone
	GI	General Infirmary, gastrointestinal
	GIT	gastrointestinal tract
	GMC	General Medical Council
	GOC	General Optical Council
	GOK	God only knows
	GP, gp	General Practitioner, group (eg blood group)
	GSF	green stick fracture
	GSW	gun shot wound
	GSWA	gun shot wound to the abdomen
	gt	gutta (drop)
	GTN	glyceryl trinitrate (a treatment for angina)
	GU, GUD	gastric ulcer, genitourinary Department
	GUM	genitourinary medicine
	GUT	genitourinary tract
	Gyn, gynae	gynaecology
H	H, h	hand, height, hormone, Hospital, hour, human
	H&C	hot and cold
	H & S	haemorrhage and shock
	HAV	hepatitis A virus
	HAVS	hand-arm vibration syndrome (also known as vibration white finger)
	Hb	haemoglobin (low in anaemia)
	Hct	haematocrit
	HDCT	high definition CT scan
	hep B, HBV	hepatitis B infection / vaccine
	hepar	of the liver
	hge	haemorrhage
	HHV	human herpes virus
	HI	head injury, hearing impaired
	HIV	human immunodeficiency virus
	HIO	head injury observations
	HO	House Officer (junior doctor)
	H/O	history of...
	HOH	hard of hearing
	hosp	hospital
	HP	House Physician (junior doctor)
	HPC	history of present condition
	HPE	history and physical examination
	HPI	history of present illness
	HPV	human papilloma virus (the cause of warts)
	HR	heart rate
	hrs, hrly	hours, hourly

H_2RA	histamine 2 receptor antagonist (a class of drugs used to treat stomach and duodenal ulcers and dyspepsia)
HRCT	high resolution CT scan
HRT	hormone replacement therapy
$5HT_1A$	5 hydroxy-trypatamine 1 agonist (a class of drug used to treat migraine)
$5HT_3A$	5 hydroxy-trypatamine 3 antagonist (a class of drug used to treat nausea)
HS	House Surgeon (junior doctor), heart sounds
HSE	Health & Safety Executive
HSV	herpes simplex virus (eg cold sore, genital herpes)
Ht	heart, height
HV	Health Visitor
HVS	high vaginal swab (taken for culture)
HW	house wife
HWB	hot water bottle
Hx	history of illness or injury
hypo	low (eg low blood sugar in diabetes, hypoglycaemia)
HZ	herpes zoster virus (eg shingles)

I

I, i	impression, Infirmary, Iodine, 1
IARC	International Agency for Research on Cancer
I & D	incision and drainage (usually of an abscess)
IB	invalidity benefit
IBD	inflammatory bowel disease
IBS	irritable bowel syndrome
I/C	sexual intercourse, in charge
ICU	Intensive Care Unit
ICD-10	International Classification of Diseases
ict	icterus, icteric (jaundice, jaundiced)
ID	identification
id	intradermal (injection within the skin)
IDDM	insulin dependent diabetes mellitus
IDLH	immediate danger to life and health
IF	internal fixation
IHD	ischaemic heart disease
ILO	International Labour Organisation
im	intramuscular (injection within muscle)
IMP	impression
in	inch
inf, Inf	inferior, infant, infection, infusion
ing	inguinal
inh, INH	inhale, inhaler
inj	injection, injury
inoc	inoculation

	insp	inspiration
	insuf	insufflate (pump air in)
	INV	investigations
	int	internal, interval, intestinal, intermittent
	ip, ip jt	intraperitoneal, interphalangeal joint of finger/toe
	IPD	inflammatory pelvic disease
	IP	In Patient
	IQ	intelligence quotient
	irr, irreg	irregular
	IT	intrathecal, intrathoracic, intratracheal
	ITU	Intensive Therapy Unit
	IUCD	intrauterine contraceptive device
	IUD	intrauterine device, intrauterine death
	iv	intravenous, intravascular, intraventricular
	IVC	inferior vena cava (blood vessel) intravenous catheter
	IVF	intravenous fluids, in-vitro fertilization
	IVU, IVP	intravenous utherogram/pyelogram (kidney and bladder X-ray)
J	jn, jx	junction
	jnt, jt	joint
	JVP	jugular venous pressure
K	K, k	kidney, knee, vitamin K, Potassium
	KDA	known drug allergies
	KJ, KJ+, KJ–	knee jerk, knee jerk present/absent
	KK_o, K_o	kidney(s) not palpable
	Kn	knee
	KO	knock out, keep open, keep on
	KT, KTx	kidney transplant
	KUB	x-ray examination of kidneys, uereters, bladder
L	L, 1	letter sent by GP, left, leg, liver, lumbar, lymph
	L1, L2, L3 etc	lumbar vertebra, lumbar nerve
	L/3	lower third
	LA	local anaesthetic, long acting, left arm
	LA1 %	local anaesthetic with 1% lignocaine
	lab	laboratory
	lacn	laceration
	lap	laparotomy (opening of the abdominal cavity)
	lat	lateral
	LBP, LBS	low back pain, low back syndrome/strain
	lg	large
	LE	left eye/ear, lupus syndrome
	Le Fort	a system of classification of fractures of facial bones (I, II or III)
	$L_{EP,d}$	a Leq for an eight-hour day (see Leq)

Leq	equivalent continuous sound level (a measure of noise exposure or dose)
LFH	left femoral hernia
LFT's	liver function tests, lung function tests
LH	luteinising hormone
LHRH	luteinising hormone releasing hormone
LIF	left iliac fossa
LIH	left inguinal hernia
linct	linctus
LIQ	left inner quadrant (eg of breast)
LJM	limited joint movement
LK	left kidney/knee
$LKKS_o$	liver kidneys and spleen not palpable
LL	left leg/lung
LMP	last menstrual period
LN	lymph node
LOC	loss of consciousness
LOM	limitation of movement
LOQ	left outer quadrant (eg of breast)
LP	lumbar puncture
LVF	left ventricular failure of the heart

M

M, m	male, malignant, maximum, mean, medical, Member, metre, minimum
M/3	middle third
MAOI's	monoamine oxidase inhibitors (antidepressant drugs)
Mag tricil	Magnesium tricilicate (an antacid)
mane	morning
mand	mandible
manip	manipulation
max	maximum
MBC	Medical Boarding Centre
MC	metacarpal
MCA	Medicines Control Agency
MC-CJ	metacarpo-carpal joint of hand
MCPJ	metacarpo-phalangeal joint of finger
MCV	mean cell volume
MDA	Medical Devices Agency
MDSO	mentally disturbed sex offender
MED 3,5 etc	medical certificates
MI	myocardial infarction
mid	middle
min	minimum
mict	micturition (passing urine)
mist	liquid mixture
M_gS0_4	magnesium sulphate (an antacid for indigestion)

ME	meningoencephalitis, middle ear
MH	medical history
ML	mid line
MMR	measles, mumps, rubella (German measles) vaccine
MND	motor neuron disease
MO	medical officer
MOPD	Medical Outpatient Department
MPJ	metacarpo-phalangeal joint of finger
MR or MRI	magnetic resonance imaging scan
MRSA	methicillin resistant staph aureus infection (or multi-resistant)
MS	multiple sclerosis, Munchausen's syndrome, musculoskeletal, Medical Student
MSU	midstream specimen of urine
MST	a morphine salt used for relief of severe pain
MSW	Medical Social Worker
MTPJ	metatarso-phalangeal joint of the foot
mvt	movement

N

N, n	nausea, nasal, nerve, Nitrogen, normal, number
NA	Narcotics Anonymous
N/A	not applicable
NAD	nothing abnormal demonstrated, no active disease, no apparent distress
N & V	nausea and vomiting
na dressing	non-adhesive dressing
naso	of the nose
NBI	no bony injury
NBM	nil by mouth, normal bowel movement
NBS	no bowel sounds
NED	no evidence of disease
neg	negative
neuro	of the neurological system, nerves
NFA	no further attendance, no fixed abode
NFCE	near fatal choking episode
NFR	not for resuscitation
NG tube	nasogastric tube
NGH	Newcastle/Notttingham etc General Hospital
NIDDM	non-insulin dependent diabetes mellitus
NIHL	noise-induced hearing loss
NK	not known
NMTBD	no more to be done
No	number
nocte	at night
NOK	next of kin
No Rx	no treatment
NP	new patient

NPhx	nasopharynx
NPL	tables of total noise immission levels (see Leq)
NPO	nil by mouth (per orum)
NR	nerve root, normal range, do not repeat
NS, N/S	nervous system, non-smoker, Nursing Sister, normal saline
NSAID's	non-steroidal anti-inflammatory drugs
NSD	no significant difference/disease/defect
NSU	non-specific urethritis
NSR	no sign of recurrence, nasal septum reconstruction, not seen regularly
NVD	nausea, vomiting and diarrhoea
NVL	no visible lesion
NWB	non-weight bearing
NYD	not yet diagnosed

O

O,o	observations, blood group O, Oxygen, pint, zero, no abnormality detected
0_2	oxygen
OA, oa	osteoarthritis, old age
O&A	observation and assessment
O&G	obstetrics and gynaecology
OAD	obstructive airways disease
obs	observations, obstetrics, obesity
OC	oral contraceptive
occ	occasional, occupation
OCD	obsessive compulsive disorder
od	omni die (once daily), overdose
ODA	Operating Department Assistant
O/E	on examination
om	omni mane (in the morning)
on	omni nocte (at night)
ONIHL	occupational noise-induced hearing loss
OPA	out patient appointment
OPD	Out Patient Department
opn, op	operation
OPRA	Occupational Physicians Reporting Activity
OPV	oral polio vaccine
OR	operating room
OTC	over the counter (drugs)
oto	of the ear, otology
out pts	out patients
ow	once weekly

P

P, p	attended GP surgery for prescription, para (number of live births), pulse rate, protein
P/3	proximal third

PA	postero-anterior (usual chest x-ray view), pernicious anaemia
palp	palpable
Pap sm	Papanicolaou (cervical) smear
Para	number of live births
path	Pathology Department/report
PCB	post-coital bleeding
PCG	Primary Care Group
PCT	Primary Care Trust
PCV	Packed cell volume
PE	pleural effusion, pulmonary embolus
PEBP	patient escorted by police
perf	perforated
PER(R)LA	pupils equal (round) and react to light and accommodation
pess	pessary
PF, PF(E)R	peak flow, peak flow (expiratory) rate (maximum breath velocity, in a short sharp expiration)
PFT's	pulmonary function tests
PGH	previous gynaecological history
PH, PHx	past history
Physio, physiox	physiotherapy
PID	prolapsed intervertebral disc, pelvic inflammatory disease
Pl	platelets (in blood), plantar
PM	post mortem
PMO	principal medical officer
PMT	premenstrual tension
PN	Practice Nurse, post-natal
PNL	peripheral nerve lesion
PNS	post nasal space
PNTh	pneumothorax (air in the pleural space)
po	by mouth
POD	prescription only drug
POP	plaster of Paris
post-	after (eg postoperative)
pr	per rectum, rectal examination
PP, Px	proximal phalanx
PPPBL	peripheral pulses palpable in both legs
PRICES	protection, rest, ice, compression, elevation, support (treatment for a soft tissue injury)
prn	pro re nata (as required)
proct	proctology (of the rectum)
prot, P+	protein, protein present in urine
prox	proximal
PRx	prognosis

	pre-	before (eg preoperative)
	Pt	patient
	PTSD	post traumatic stress disorder
	PUO	pyrexia of unknown origin
	PUVA	psoralen + ultra-violet light (treatment for psoriasis)
	pule	powder
	pv, PV	per vaginum, vaginal examination, plasma volume
	PVD	patient very distressed
	PVS	persistent vegetative state
	Px	phalanx
Q	q	every, each
	QA	quality assurance
	qds	quater die sumendus (four times daily)
	QEH	Queen Elizabeth Hospital
	qh, gd, gn	every hour/day/night
	q2h, q3h	every 2 hours, every 3 hours
	qoh,qod,qon	every other hour/day/night
	qqh	quarta quaque hora (every four hours)
	QOL	quality of life
R	R,r	race, rate, rectal, referred, right
	RA	rheumatoid arthritis, right arm
	RADS	reactive airways disease syndrome
	RAST	radio-allergo sorbent tests (a test for allergic antibodies in blood)
	R&R	rate and rhythm
	radiol	Radiology Department/report
	RAP	recurrent abdominal pain
	RBC	red blood cells
	RBS	random blood sugar
	RCN	Royal College of Nursing
	Re, RE	regarding, right eye/ear
	recur	recurrent
	Reg, reg	Registrar (junior doctor in specialist training), regular
	rehab	rehabilitation
	rep, rept	repeat
	REM	rapid eye movement (a phase of sleep)
	RF	renal failure
	RhA	rheumatoid arthritis
	Rh+ and Rh–	blood groups Rh+, Rh–
	Rh fever	rheumatic fever
	RI	Royal Infirmary
	RIDDOR	Reporting of Injuries, Diseases and Dangerous Occurrences Regulations
	RIF	right iliac fossa

RIP	died
RL	right leg/lung
RMF	right middle finger
ROM	range/restriction of movement
rotn	rotation
RP	radial pulse
rpt	repeat
RRR	regular rhythm and rate
RS, RS_o	respiratory system, respiratory system normal
RSM	Royal Society of Medicine
RT	radiotherapy
RTA	road traffic accident
RTI	respiratory tract infection
RTW	return to work
RUA	reduced under anaesthetic (eg dislocation)
Rx	prescription, treatment

S

S,s	single, spleen, surgery
S 1, S2, S3 etc	sacral vertebra or nerves
S1--S2+0	heart sounds normal with no added sounds
SA	sexual assault, sexual addict
SAD	seasonal affective disorder
SAC	sacrum, short arm plaster cast
SB	small bowel, stillbirth, spina bifida
SB	seen by
SBM	small bowel barium X-ray
SBS	shaken baby syndrome
SC	spinal cord, sugarcoated (eg tablets)
scap	scapular
SCC	spinal cord compression, squamous cell carcinoma (skin cancer)
SCBU	Special Care Baby Unit
SCP	Society of Chirpodists & Podiatrists
SCU	Self Care Unit
SDH	subdural haematoma
SH, sh	social history, sexual harassment, shoulder
SHI	significant head injury
SHO	Senior House Officer (junior doctor)
SI	sacroiliac joint, sexual intercourse
sibs	siblings
SID	sudden infant death
SII, SIW	self inflicted injury/wound
SIJ	sacroiliac joint
sk tr	skeletal traction
SLR, SLRT	straight leg raising test
SLWC	short leg walking cast
SMO	senior medical officer

	SMR (R)	sub mucosal resection of nose (+ rhinoplasty)
	SOA	swelling of ankles
	SOB	short of breath
	sol, SOL	solution, space occupying lesion eg brain tumour
	SOPD	Surgical Outpatient Department
	sp cd	spinal cord
	spont	spontaneous
	spr	sprain
	Sp Reg	Specialist Registrar
	SR	Senior Registrar (a senior junior doctor)
	S/R, sr	slow release drug
	SRF	severe renal failure
	SSDA	Social Security Disability Allowance
	SSRI	selective serotonin re-uptake inhibitor (a class of anti-depressants)
	staph	staphylococcal infection
	stat	immediately
	STDs	sexually transmitted diseases
	STG	split thickness graft
	STLOM	swelling, tenderness and limitation of movement
	STSG	split thickness skin graft
	strep	streptococcal infection
	sub-	below
	subac	subacute
	subling	under the tongue
	sublux	displaced below
	sup	supine
	suppos	suppository
	supra-	above
	SUUD	sudden unexpected unexplained death
	SVT	supraventricular tachycardia (an abnormal heart rhythm)
	SWORD	Surveillance of Work Related and Occupational Respiratory Diseases
	Sx	symptoms, suction
	synd	syndrome
	syr	syrup
	Syst	system
T	T, t	tablets, temperature, time, toe, total, treatment
	T1, T2, T3 etc	thoracic vertebra or nerve
	T4	thyroid function test
	TA	traffic accident
	T/A	time and amount
	TAb	therapeutic abortion
	tabs	tablets

	tachy	rapid eg tachycardia (rapid pulse) or tachypnoea (rapid breathing)
	T&E	trial and error
	TAH	total abdominal hysterectomy
	TAT	tetanus antitoxin
	TATT	tired all the time
	TB	tuberculosis
	TBI	traumatic brain injury
	TCA	to come again
	TCABG	triple coronary artery bypass graft
	TCI	to come in
	tds	ter die sumendus (three times daily)
	T/D	treatment discontinued
	TENS	transcutaneous electrical nerve stimulation (a physical method of pain relief)
	TFT's	thyroid function tests
	TH	tension headache
	THR	total hip replacement
	TI	thoracic injury
	TIA	transient ischaemic attack (mini-stroke)
	tib and fib	tibia and fibula
	TICTAC	computer aided tablet and capsule recognition system used in overdose
	TG	tendon graft
	TIPJ	terminal interphalangeal joint of toe or finger
	TKR	total knee replacement
	TLC	tender loving care
	TMJ	temporo-mandibular joint
	TNTC	too numerous to count
	To	body temperature
	TOD	took own discharge
	TOP	termination of pregnancy
	TPN	total parenteral nutrition
	TPR	temperature pulse and respiration
	tr	trace, treatment
	TSH	thyroid stimulating hormone
	TT	tetanus injection
	TURP	transurethral prostatectomy
	trop	tropical
	Ts & As	tonsils and adenoids
	TTOs	to take out of hospital (eg drugs)
U	U, u	unit, upper, urine
	U/2, U/3	upper half, upper third
	UA	urine analysis
	U & E	urea and electrolytes (a biochemical test of salt balance and kidney function)

	UDC	usual disease of childhood
	UGI	upper gastro-intestinal tract
	UIQ	upper inner quadrant (of breast, abdomen etc)
	UKCC	UK Central Council for Nursing, Midwifery and Health Visiting
	unilat	unilateral
	ung	unguent (ointment), ungual (relating to nails)
	UOQ	upper outer quadrant (of breast, abdomen etc)
	urol	urology
	URT	upper respiratory tract
	URTI	upper respiratory tract infection
	U S	ultrasound examination
	UTI	urinary tract infection
V	V, v	vein, vertebra, virus, visit, volume, vomiting, 5
	VA	visual acuity
	vacc	vaccination
	vag spec	vaginal speculum
	VD	venereal disease
	VF	ventricular fibrillation (an abnormal heart rhythm associated with sudden death)
	vg (h)	very good (health)
	VI	virgin, 6
	VIF	visual fields intact
	VIWF	vibration induced white finger
	vit	vitamin
	vs, VS	see above, vital signs
	VT	ventricular tachycardia (an abnormal heart rhythm)
	VV	varicose veins
	VWF	vibration white finger
W	W, w	ward, water, week, weight, white, widowed, wife, with
	WBC, WCC	white cell count
	WC	whooping cough
	Wd	ward
	Wd Rnd	ward round
	WDWN	well developed and well nourished
	WE	entry wound
	WIG	Western Infirmary Glasgow
	Wk	week
	WL	waiting list
	WLE	wide local excision
	WNF	well nourished female
	WNM	well nourished male
	wo	without
	wr, WR	wrist, Wassermann reaction (test for syphilis)

	WRULD	work related upper limb disorder (the modern and preferred term for "RSI")
	wt	weight
	WX	exit wound
X	X, x	except, multiplication, magnification, 10
	X match, XM	cross match blood
	XR	X-ray examination
	Xs	excess
Y	y	year
	yd	yard
	YLC	youngest living child
	y o	years old
	YOB	year of birth

Some common medical symbols

1/7	one day
1/52	one week
1/12	one month
+	positive (present)
++/+++	positive (present) to an abnormal degree
–	negative (absent)
↑	increased (up)
↑	decreased (down)
≠	not equal
>	greater than
<	less than
→	to (in the direction of)
∅	normal
#	fracture
Rx	treatment or therapy
†	deceased
Δ	diagnosis

APPENDIX 2
MEDICAL GLOSSARY

A

Abduction	Moving a limb outwards from the medial line
Ablation	Cutting away tissue or abnormal growth
Abscess	A circumscribed collection of pus
Acetabulum	Socket on the lower part of the pelvic bone in which the head of the femur (hip bone) is situated, and attached by ligaments
Achilles tendon	The tendon of the calf muscle, which inserts into the calcaneum bone (heel)
Acrocyanosis	Slow blood circulation causing bluish discolouration of hands and feet
Acromio-clavicular Joint	The joint between the acromion process of the scapula and the lateral end of the clavicle (collar bone)
Acroparaesthesia	An unpleasant tingling sensation in hands and feet
Acute	Severe symptoms, brief in duration
Adaptation	Normal reflex elevation of hearing threshold in response to loud noise
Adduction	The drawing of a limb towards the medial line
Adenitis	Inflammation of a gland
Adenopathy	A diseased gland
Adventitia	Loose connective tissue on the outer surface of a blood vessel
Agnosia	Inability to recognize familiar auditory, tactile or visual stimuli, in spite of normal sensation
Agoraphobia	Fear of open spaces (the opposite of claustrophobia)

Agraphia	Inability to express thoughts in writing
Air conduction	The transmission of sound through the outer and middle ear to the internal ear
Akinesia	Loss of muscular response
Alexia	Difficulty in reading
Allergic reaction	A hypersensitive immune reaction to drugs and other biological materials (eg pollen, bee venom, nuts) following initial exposure
Alveolar fracture	A fracture involving the alveolus
Alveolus	The part of the upper or lower jawbone that holds the roots of the teeth. A small sac where gas exchange takes place in the lung.
Amnesia	Inability to recall information following trauma. May be for events prior to trauma (retrograde), or subsequent events (anterograde), or both
Anastamosis	Joining up two ends of a hollow organ
Amputation	Removal of a portion of a limb or digit
Analgesia	Pain relieving medication
Angina	Tight constricting chest pain due to reversible insufficiency of the coronary arterial blood supply to the myocardium (heart muscle)
Ankylosis	The replacement of a joint by fibrous tissue eg following infection
Anodyne	Treatment to ease pain
Anosmia	Loss of sense of smell
Anoxia	Insufficient oxygen supply to the brain, which if not reversed may cause anoxic brain damage
Anterior	Front part of the body or any bodily structure
Anticonvulsant	Drug used to suppress epileptic seizures
Antidepressant	Drug used to treat depressive illness
Anti-inflammatory	Usually refers to drugs used to treat pain and inflammation not associated with infection, of which the most common group are the non-steroidal anti-inflammatory drugs (NSAID)
Antipyretic	Fever reducing drug

Antrectomy	Operation for the removal of part of the stomach
Anuria	Failure of secretion of urine; a symptom of kidney failure
Aphasia	Absence of ability to speak
Apraxia	Loss of the ability to perform skilled voluntary movements in the absence of muscle paralysis
Arachnoid mater	One of the membranes (the meninges) lining the brain and spinal cord. The sub-arachnoid space within contains the cerebro-spinal fluid
Arteriole Artery	A small artery blood vessel conveying oxygenated blood
Arthritis	Inflammation of a joint
Arthrodesis	Operation to fuse a joint
Arthro-	Prefix meaning "relating to a joint"
Arthropathy	Disease of a joint
Arthroplasty	Re-fashioning of a diseased joint by surgery
Arthroscopy	Examination of a joint with an endoscope
Arthrotomy	Surgical opening of a joint
Asbestos bodies	Coated fibres of asbestos, seen on light microscopy of lung tissue
Asbestosis	Form of pneumoconiosis caused by inhalation of asbestos
Aseptic necrosis	Bone death without infection
Asphyxia	Impaired or absent ventilation of the lungs
Aspiration	Withdrawal of fluid from a body cavity by means of suction
Asthenia	Being weak
Asthma	A lung disease characterised by hyperactive airways and reversible airflow obstruction
Ataxia	Loss of coordination of the head, trunk or limbs
Atheroma	Degeneration of artery walls with fatty deposits, sometimes leading to thrombosis

	Atherosclerosis	Disease of arteries in which calcarious plaques develop; hardening of the arteries
	Athetoid	Slow writhing movements associated with certain forms of brain damage, including athetoid cerebral palsy
	Atrophy	Wasting away of tissues through cell degeneration
	Audiogram	A graph recording level of hearing, at different frequencies through tests carried out on an audiometer
	Audiometry	Assessment of hearing function
	Auditory ossicles	Bones of middle ear (hammer, anvil and stirrup)
	Auricle	Outer ear
	Autonomic nervous system	The involuntary nervous system, concerned with regulation of the circulation, breathing and other basic functions
	Avascular necrosis	Death of bone or tissue through interruption of blood supply
	Avulsion	A tearing, such as of skin, ligaments or bone from adjacent structures
	Avulsion fracture	The avulsion of a fragment of bone due to stress on the attached ligament or tendon
	Axilla	The armpit
	Axon	A nerve fibre
B	Babinski reflex	Abnormal extension of the great toe when the sole of the foot is scratched. May occur when higher control of muscle function is lost, for example following a stroke
	Baritosis	A form of lung disease (from barium dust)
	Barotitis	Earache when eg climbing, diving or descending in an aeroplane
	Barotrauma	Pressure injury to the middle ear after eg climbing, diving or descending in an aeroplane
	Basal ganglia	A cluster of nerve cells (grey matter) situated deep in the cerebral hemispheres, and concerned with regulation of motor and sensory impulses. They may be injured in

	cerebral palsy, and are diseased in Parkinson's disease
"Beat" conditions	Sterile subcutaneous cellulitis affecting hand, elbow or knee
Behaviour therapy	A form of psychotherapy directed to regulating abnormal behaviour
Berylliosis	A pneumoconiosis due to beryllium dust
Bilateral	Occurring on both sides
Binocular	Involving both eyes
Biopsy	Removal and examination of tissue for diagnostic purposes
Blepharitis	Inflammation of the eyelids
Bone conduction	The transmission of sound to the internal ear through vibration of cranial bones
Bone graft	Transplantation of bone from one site to another
Bowel	The intestine
Brace	A removable splint or support
Brachial plexus	A junction of nerve roots at the root of the neck formed from the 5th to 8th cervical nerve roots and the first thoracic nerve root which branch to supply the muscles of the upper limbs
Bradykinesia	Abnormally slow movements, a symptom of Parkinson's disease
Brain stem	The lower posterior part of the brain immediately above the spinal cord, which contains vital centres controlling breathing and heart function. Serious brain stem injury or death is incompatible with life
Break	A fracture
Broca's area	The area of the dominant cerebral cortex associated with speech. Injury or disease may result in aphasia or dysphasia
Bronchi	The branching air passages which lead from the trachea (windpipe) to the areas of gas exchange in the lungs (the alveoli)

Bronchial	Of the bronchi
Bronchioles	The smallest bronchi
Bronchoscope	An instrument to examine the bronchi, which may be either rigid or flexible
Bruise	Soft tissue trauma associated with bleeding into the tissues
Bucket handle tear	The typical appearance of a longitudinal tear of the semi-lunar cartilages (the menisci) of the knee
Bulbar	Relating to the part of the brain stem known as the medulla
Bulbar palsy	Disease or injury of the bulbar part of the brainstem, which results in difficulty swallowing
Burr hole	A hole drilled into the skull, for example to relieve pressure caused by bleeding
Butterfly fragment	A curved triangular bone fragment seen in some tibial fractures
Bursa	Natural pouches of fibrous tissue lined by smooth cells and containing a little fluid. Situated at points where there is pressure or friction, they allow free movement of skin over joints
Bursitis	Inflammation of a bursa (eg housemaid's knee)
Byssinosis	A form of lung disease, resulting from inhalation of cotton flax or hemp dust

C

Cacosmia	Irregularity in the sense of smell caused by brain damage
Calcaneovalgus	A foot deformity (upwards and outwards)
Calcaneus	A foot deformity (upwards but not outwards)
Callus	Bony material growing around and between two ends of a fractured bone in healing
Canthus	The corner eg of the eye
Capsulitis	Inflammation about a joint ie in the joint capsule
Cardiac	Of the heart

Caries	Tooth or bone decay
Carpal	Of the wrist
Carpal tunnel	Space formed by the bones of the wrist on the posterior surface and a relatively inelastic membrane (flexor retinaculum) on the anterior surface which conducts the median nerve
Carpal tunnel Syndrome	Entrapment of the median nerve at the carpal tunnel of the wrist
Carpus	Bones of wrist
Catharsis	Cleansing the bowels; purgation; the relief of strong feelings or tension
Causalgia	Intense continuous burning pain, usually following nerve injury
Cephalic	Of the head
Cerebrovascular accident	Stroke
Cervical	Of the neck
Cervical spondylosis	Arthritic degeneration of bones and joints of the neck
Cervical syndrome	A syndrome of pain, aching and stiffness in the neck, sometimes associated with prolonged static neck flexion
Cholecystectomy	Removal of gall bladder
Chronic	Slowly developing, long lasting
Chronic bonchitis	A lung disease associated with fixed airway obstruction and excess mucus production
Cicatrix	The scar of a healed wound
Circumduction	Moving a limb in a circular movement
Claudication	Muscular pain in the lower limbs on exercise due to insufficiency of circulation
Closed fracture	Fracture with intact skin
Clubbing	Bulbous swelling of nail pulps seen in chronic lung and other diseases.
Coccydynia	Pain in the coccyx and surrounding area
Coccyx	Vestigial human tail; small triangular bone attached to the lower part of the sacrum

Cochlea	Part of inner ear. A snail-like structure containing organ of hearing
Colles' fracture	Fracture across the lower end of the radius
Colon	Large intestine
Colitis	Inflammation of the colon
Colposcopy	Examination of the cervix through a microscope
Comminuted	Bone fractured into several pieces
Compound fracture	A fracture in which the skin is broken (the modern term is "open")
Condyle	Rounded bulge at the end of some bones
Congenital	Present at birth
Contralateral	On the opposite side of the body (as opposed to ipsilateral, on the same side)
Contusion	A bruise
Cordectomy	The operation to remove a vocal cord
Coronoid process	The protuberance towards the top of the ulna
Cortex	Outer part of an organ or bone
Costal	Of the ribs
Coxa	Hip
Coxarthrosis	Another expression for osteoarthritis of the hip (usage mainly European)
Crepitations	Crackling noises caused by congestion or inflammation of the lungs
Crepitus	Grating caused by bone running against bone or roughened cartilage (usually associated with movement of arthritic joint)
Crural	Of the leg or thigh
Cubitus	Elbow
Cyanosis	Blueness of skin owing to circulation of inadequately oxygenated blood
Cyclitis	A form of inflammation of the eye
Cycloplegia	A form of paralysis of the eye muscle

Cystic duct	The normal drainage channel for the gall bladder into the upper bowel
Cytology	Examination of cells, usually for signs of malignancy

D

Dactylitis	Inflammation of a digit (eg toe, finger)
Daltonism	Loosely used to mean colour blindness. Accurate meaning: inability to distinguish between red and green
Debridement	The removal of foreign matter in a wound (such as dirt, clothing, dead tissue, etc) to clean the wound and facilitate healing
Decubitus ulcer	Bedsore
Defibrillation	Restoration of heart rhythm by means of an electric shock
Delayed union	A fracture complication in which the bone ends heal very slowly
De Quervain's Syndrome	Pain, discomfort and occasional clicking involving two of the tendons of the thumb, abductor pollicis longus and extensor pollicis brevis
Dermal	Of the skin
Dermatitis	Inflammation of the skin
Dermatoplasty	General generic term covering skin grafting, plastic surgery etc
Devitalize	To remove the "nerve" of a tooth or an organ
Diaclasia	The deliberate surgical fracture of a bone to reset deformity
Dialysis	The removal of harmful toxins from the blood by an artificial kidney
Diaphoresis	Increased perspiration
Diaphysis	The central shaft of a large bone
Diffuse pleural thickening	Progressive thickening of panetal pleura of lung, a condition associated with asbestos exposure
Diplegia	Paralysis of both lower limbs
Diplopia	Double vision

Disarticulation	To separate joints (by injury or surgery)
Dislocation	Separation of the components of a joint without fracture
Distal	Extremity of a limb or organ; furthest from centre of body
Dorsal spine	Part of the spine where ribs join
Dorsiflexion	Backward movement of joints
Dorsum	Back or outer surface
Dys-	(prefix) Difficult, abnormal, impaired
Dysarthria	Difficulty in getting tongue round words – slurred speech
Dysphasia	Partial inability to speak
Dyspnoea	Difficulty in breathing, breathlessness
Dystocia	Difficult childbirth
Dysuria	Pain during urination

E

Eburnation	The final result of osteoarthritis, when the cartilage at the articulating surface of a bone is totally worn down
Eclampsia	A type of epileptic convulsion, to which pregnant women are particularly susceptible
Ecdysis	The shedding of skin
Eczma	A type of inflammation of the skin
Edentulous	No teeth
Effusion	The escape of fluid (eg pus, blood etc) into a body cavity
Electro-	
Encephalogram	A graph derived from a recording of brain electrical activity, used in the diagnosis of epilepsy
Embolism	Blockage of blood vessels by material (such as blood clots) carried through larger blood vessels in the blood stream
Embrasure	The gaps existing between neighbouring teeth

Eminence	A projection – usually describing that on a bone
Emphysema	Collection of air in certain parts of the body where it is not usually present; loss of the supporting tissue of the lung
Endometrium	Lining of the womb
Enarthrosis	A joint, such as the hip or shoulder, which operates by way of a ball and socket
Endothelium	Internal lining of blood vessels
Enterocentesis	The removal of an undue quantity of fluid from the stomach (or intestine) by use of a hollow needle
Enucleate	To remove an eye
Enuresis	Bed wetting
Epicondyle	Protuberance above a condyle at the end of a bone which articulates with another bone
Epicondylitis	Inflammation in region of epicondyle eg tennis elbow, at the site of attachment of extensor tendons of the forearm to lateral epicondyle; golfer's elbow, at the site of attachment of the flexor tendons to medial epicondyle
Epicranius	The scalp muscle
Epidermis	Outer layer of skin
Epidural	Relating to the outermost membrane covering the spinal cord
Epiphysis	The growing part of a bone at the end of the bone
Epiphysitis	Inflammation of the epiphysis
Epistaxis	Nose bleed
Equinovarus	A foot deformity (downwards and inwards)
Equinus	A foot deformity (downwards)
Erythema	Superficial redress of skin
Eustachian tube	Passage connecting middle ear to pharynx
Eutocia	Normal labour and childbirth without complications

	Eversion	The movement of the subtalar joint in which the foot is turned upwards and outwards (opposite of inversion)
	Excise	To cut tissue away surgically
	Excision Arthroplasty	An arthoplasty in which the joint is also excised
	Exsanguinated	Blood loss to a dangerous degree
	Extension	Straightening out of a joint
	External auditory meatus	Passage leading from outer to middle ear
F	Facet	A surface on a bone which is small and flat
	Fatigue fracture	A fracture due to repeated minor stress
	Fauces	Structures at the back of the throat
	Fibrosis	Thickening of tissue due to scarring
	Fistula	Abnormal communication
	Flaccid	Muscles with no tone
	Flexion	Bending of a joint
	Flexor	The bending movement of a joint
	Forward angulation	A deformity at fracture sites
	Fracture dislocation	A dislocation accompanied by a fracture
	Fusion	The joining together, normally of two bones
G	Gastrotomy	The surgical opening of the stomach
	Genioplasty	Plastic surgery to build up cheek bone
	Genu	The knee
	Gibbus	An angular and sharp backward curvature of the spine, usually the result of disease or tumour
	Gingivitis	Gum inflammation
	Ginglymus	A hinge joint
	Gluteal	Of the buttock
	Gonarthrosis	Osteoarthritis of the knee (usage mainly European)
	Gustatory	Relating to taste organs

H	Haemarthrosis	The swelling and pain resulting from bleeding into a joint
	Haematoma	Collection of blood forming swelling
	Haemo-	
	Pneumothorax	Blood and air in the pleural cavity (which sometimes has to be drained)
	Halitosis	Bad breath
	Hallux	Great toe
	Hamstring	A tendon at the back of the knee
	Hearing threshold	
	Level	The threshold of hearing measured in dB
	Hemiplegia	Paralysis of one side of the body
	Heparinised	The effect of the use of a drug to prevent clotting
	Hepatiscostomy	The making of a surgical opening in the main bile duct
	Hernioplasty	The surgical repair of a hernia
	High fence	Level at which hearing disability is total (see low fence)
	Histogenesis	Tissue formation
	Histology	Study of structure of tissues by means of special staining techniques together with light and electron microscopy
	Homeostasis	The system whereby body functions (temperature, blood pressure, etc) remain in equilibrium whatever the outside environment
	Homoplasty	Repair of damaged tissue by grafting on similar tissues from another of the same species
	Hydramnios	Excessive amniotic fluid
	Hydrocephalus	Distension of the cerebral ventricles by the accumulation of fluid
	Hyper-	(prefix) Over, to excess, above normal
	Hyperacusis	Excessive sensitivity to sight, sound, taste and smell
	Hyperaemia	Abnormally increased blood flow

Hyperaesthesia	Excessive sensitivity to sound, taste, smell or feeling
Hyperextension	Abnormal movement of a limb beyond its limit
Hypertrophy	Enlargement of cells with resultant increase in size of organ or tissue
Hyperventilation	Abnormally fast or deep breathing
-Hypo-	(prefix) Below normal, decreased
Hypocalcaemia	Metabolic disturbance with a drop in the serum level of calcium
Hypohidrosis	Failure-to sweat sufficiently
Hypoplasia	Failure of tissue to develop fully
Hypotropia	Squint
Hysterectomy	Removal of the womb (uterus) May be removed via vagina or abdomen

I

Ileum	Small intestine
Ilium	Superior bone of pelvis (hip bone)
Implant	In dentistry, the plate to which false teeth are attached
Induration	Excessive toughening of an organ or tissue
Infarction	Death of whole or part of organ, caused by obstruction of relevant artery (eg myocardial infarction)
Inguinal	Of the groin
Intercostal	Situated between ribs
Internal rotation	Rotating a limb inwards
Inter-vertebral disc	Disc-like flattened cartilaginous structure between two vertebrae
Intra-articular	Within the synovial cavity of a joint
Intracranial	Inside the skull
Intramedullary	Within the medullary cavity of a long bone
Intravenous injection	An injection into the vein
Inversion	Downwards and inwards movement of the foot at the sub-talar joint

	Ipsilateral	On the same side of the body (the opposite of contralateral)
	Iridoplegia	Paralysis of the iris in the eye.
	Iridotomy	Surgical incision of the iris
	Ischaemia	Deficiency of blood in part of body
	-itis	(suffix) Inflammation
J	Jaundice	Yellowing of skin and other tissues through accumulation of bile products in the body
K	Keloid	Overgrowth of fibrous tissue usually on site of scar of previous injury
	Keratectomy	The surgical removal of a superficial layer of the cornea
	Keratitis	Inflammation of the cornea
	Keratoplasty	Corneal graft
	Kyphoscoliosis	Spinal deformity with sideways and forward concavity
	Kyphosis	Spinal curvature with forward concavity (normal in dorsal spine)
L	Labial	Of the lips either of the mouth or the vulva
	Labioplasty	Surgical repair to the lips either of the mouth or the vulva
	Lachrymal	Relating to the production of tears
	Laminectomy	Operation to the spine to reach the discs, nerve roots or spinal cord
	Laparoscopy	Technique to inspect pelvic organs by distending abdominal cavity with gas, and then inserting a laparoscope below the umbilicus
	Laparotomy	Incision into abdomen to investigate or treat disease
	Laryngectomy	The removal of part of the larynx
	Lateral	To the side, external
	Lesion	Damage, injury, harmful change in function or texture of organ

	Leucocyte	A white blood cell
	Leucocytosis	Elevated white cell count, usually seen in infection
	Ligament	Short, flexible fibrous tissue binding together bones of the body
	Lordosis	Curvature of the spine with backwards concavity; normal in the cervical and lumbar spine (the opposite of "kyphosis")
	Low fence	Level at which elevation of hearing threshold first disabling
	Lumbago	The lower part of the back. NB not a diagnosis
	Lumbar	Applying to the lower back in the region of the lumbar spine or adjacent muscles
	Lumbosacral spine	That part of spine composed of lumbar vertebrae and sacrum
	Lumen	The cavity of an organ or blood vessel
	Lunate bone	A bone of the wrist which articulates with the radius and ulna to the rear and the capitate bone to the front
M	Macrophage	A type of white cell which plays a role in removal of dust particles from the lung
	Mal-union	Fracture which heals in an unacceptable position
	McMurray's test	Test for tear of knee cartilage
	Macrocytosis	Red blood corpuscles being larger than normal
	Mal-union	The failure of bones properly to align after fracture resulting in deformity
	Mammoplasty	Operation to reduce the size of the female breasts
	Mandible	Lower jaw
	Manipulation	The procedure whereby a joint is moved to reduce or eliminate stiffness; can be used to replace a dislocation
	Mast cells	Cells found in the airways and other tissues, concerned with mediating acute allergic reactions

Mastectomy	Removal of breast
Mastication	Chewing food
Maxilla	Either of the pair of bones constituting the upper jaw
Medial	Middle, internal
Ménière's disease	A disease of the inner ear causing acute attacks of tinnitus and vertigo
Meniscetomy	Removal of cartilage in knee (to ease pain and locking of joint)
Meniscus	Semi-lunar cartilage as in wrist or knee joints
Mesothelioma	A malignant tumour of the lining of the lung, which enlarges slowly to encase the lung. There is a gradual increase in breathlessness, pain, and fluid may build up in the pleural space. Almost invariably the result of exposure to asbestos. Rarely arises in peritoneum or pericardium
Metaphysis	That part of the bone which is growing between the ends and the shaft
Metatarsalgia	Aching pain in the region of the metatarsal bones of the foot
Metrorrhexis	Rupture of the womb
Microcytosis	Red blood corpuscles being smaller than normal
Monoplegia	Paralysis of one limb
Mucociliary epithelium	Mucus secreting cells lining lower air passages
Myectomy	The surgical removal of part of a muscle
Myelogram	X-ray of the spinal canal after injection of a radio-opaque contrast medium into the subarachoid space
Myocardium	Middle of the three layers which form the walls of the heart
Myodynia	Muscle pain
Myotomy	Operation to divide a muscle

N	Nares	Nostrils
	Nasal	Of the nose
	Necrosis	Death of tissue or bones
	Nephritis	Inflammation of kidneys
	Nephrotomy	Operation into the kidney
	Neurasthenia	Literally nervous debility. A psychological syndrome, typically with irritability, headache, dizziness, anxiety, impatience. May follow head injury or exist as a neurosis
	Neurectomy	Removal of nerve
	Neurocranium	That part of the skull in which the brain is situated
	Neuroma	Benign tumour of nerve sheath, which may complicate injury and give rise to chronic pain and sensory disturbance
	Neuropathy	Any disorder of peripheral nerves (the symptoms being usually numbness and weakness
	Neuropraxia	Temporary failure or impairment of nerve function due to a neuro-surgical condition
	Neuroretinitis	Inflammation of optic nerve and retina
	Neurosis	A pathological emotional state, eg anxiety, phobia, depression, hysteria
	Node	A small knot of tissue
	Nutation	Nodding the head
	Nystagmus	Involuntary fluttering of eyes
O	Occiput	Back of skull
	Occupational therapy	Practical exercises to facilitate recovery from illness
	Odont-	(prefix) Of a tooth
	Oedema	Swelling due to accumulation of fluid in tissue
	Oesophagitis	Inflammation of the gullet
	Olfaction	The sense of smell
	Oligo-	(Prefix) Small, few

Onych-	(Prefix) Of the nails
Ophthalmectomy	Surgical removal of eye
Ophthalmitis	Inflammation of eye
Ophthalmorrhexis	The bursting open of the eye ball
Opposition	Movement of the thumb across to meet the little finger
Organ of Corti	Organ of hearing within cochlea
Orthodontist	Specialist who corrects the malocclusion of teeth
Orthosis	An appliance, brace, splint or caliper to correct deformity or provides support
Ostectomy	Removal of bone
Osteitis	Bone inflammation
Osteoarthritis	Degenerative disease of joint with wear of joint cartilage
Osteoarthrotomy	Operation to excise bone adjoining joint
Osteochondritis	Inflammation of a bone and its adjacent cartilage
Osteomyelitis	Infection of bone
Osteoporosis	Loss of bony tissue causing bones to become softer and liable to bend or to fracture
Osteotomy	An operation to cut across bone, usually to correct deformity
Otitis	Inflammation of the ear
Otosclerosis	A fusion of the bones of the middle ear producing deafness
-otomy	(Suffix) The surgical division of a structure
Otoplasty	Surgical repair to the ear
Oval window	Membrane separating middle and inner ear

P

Paget's disease	An inflammatory bone disease, which may cause conductive deafness
Palmar	Of the palm of hand (see volar)
Palmarflexion	Bending of wrist towards arm

Palsy	Paralysis
Para-	(Prefix) Alongside of, abnormality of, beyond
Paraesthesia	Pins and needles
Paraplegia	Paralysis of both legs resulting from an injury to the dorsal or lumbar cord
Paresis	Partial paralysis
Parietal	The outer lining of the membranes of each organ in the peritoneum, pericardium and pleura
Parietal bone	One of the bones forming the top of the skull
Parosmia	Experience of a smell which isn't there
Patch testing	The application of a number of substances to the body to establish which are responsible for an allergy
Patella	Knee cap
Pathogenic	Disease-producing
Perceptual rivalry	Loss of ability to perceive sounds clearly in presence of background noise
Peri-	(Prefix) Around
Periarthritis	Inflammation of tissue around a joint
Pericardium	The membrane which encloses the heart
Periostitis	Inflammation of the membrane on surface of bone
Peritendinitis Crepitans	A mild variant of tenosynovitis, usually more proximal to the wrist, on the extensor aspect or back of forearm
Peritoneum	The membrane enclosing the abdominal organs
Phalen's test	A test for carpal tunnel syndrome (abrupt flexion or extension of wrist provoking symptoms)
Phlebitis	Inflammation of a vein
Photosensitivity	Abnormal or excessive sensitivity to light
Physiotherapy	Treatment to restore full movement of a limb, including massage, infrared/ultra-violet rays, manipulation, exercise, etc

Pinna	Outer ear
Plantar	Of sole of foot
Plantar flexion	Flexing foot
Plaque	Raised patch of skin
-plegia	(Suffix) Paralysis
Pleural biopsy	Removal of a sample of pleural tissue for diagnostic purposes
Pleural cavity	The potential space between the visceral and parietal pleura, ie the covering of the lungs (visceral pleura) and of the inner surface of chest wall (parietal pleura)
Pleural effusion	A collection of fluid between perietal and visceral pleura
Pleural plaque	A benign circumscribed area of pleural thickening associated with asbestos exposure
Pneumoconiosis	Lung disease caused by inhalation of dust
Pneumothorax	Collection of air in pleural cavity (through lesion in lung or wound in chest wall)
Pollux	Thumb
Polypus	Tumour attached by stalk to surface
Presbyacusis	Hearing loss due to aging
Priapism	Persistent erection of the penis
Proprioception	Joint position sense
Prolapse	Slipping down or displacement of organ or structure (eg prolapsed intervertebral disc)
Pronation	Rotational movement of forearm so that the palm faces downwards
Prostatectomy	Removal of the prostate gland by surgery
Prosthesis	An appliance used to replace a body part
Proximal	Nearer to centre of body
Pseudo-arthrosis	A false joint, usually resulting from the non-union of a fracture
Pulmonary	Of the lung
Purgative	Laxative

	Pyarthrosis	A pus-filled joint
	Pyrexia	Fever
Q	Quadriplegia	Weakness, at worst, paralysis, of upper limbs, trunk and legs, resulting from a high cervical cord injury. Also called tetraplegia
R	Radial	Towards the same side of the forearm as the radius (the same as "lateral", but not when forearm is pronated)
	Radius	The outer bone of the forearm
	Raynaud's Disease	A condition in which the arteries of the fingers are unduly reactive and go into spasm on exposure to cold. A primary constitutional syndrome ie with no apparent cause (compare with Raynaud's phenomenon)
	Raynaud's Phenomenon	Symptoms similar to Raynaud's disease, occurring where there is an identifiable cause eg HAVS
	Recurvatum	Abnormal hyperextension, typically of knee or elbow
	Reduction	Bringing back to normal position, usually a fracture or dislocation
	Reflex sympathetic Dystrophy	Severe spontaneous pain, like causalgia, sometimes associated with disturbances of blood flow and wasting of skin and muscle, which may follow trauma, even of minor degree, and non-traumatic inflammatory or arthritic conditions, usually in a limb or limb girdle (see causalgia and Sudek's atrophy)
	Regional fibro-myalgia	A syndrome of pain and muscular tenderness affecting upper limb and neck
	Renal	Of the kidneys
	Resection	Any surgical removal of a portion of the body
	Rheumatoid arthritis	Progressive inflammatory disease of joints with considerable disfiguration
	Right homonymous Hemianopia	Visual loss of the right field of vision
	Rigor mortis	Stiffening ensuing soon after death

	Rotation	Movement about the long axis of a limb (usually the hip or shoulder) or the spine
S	Sacrum	Curved triangular element of backbone consisting of five fused vertebrae (sacral vertebrae)
	Sacrodynia	Pain in sacrum
	Sciatica	Pain in distribution of sciatic nerve down back of leg
	Scoliosis	Spinal deformity with sideways concavity
	Sequestrum	A dead piece of bone occurring in osteomyelitis
	Sesamoid	Bones lying within tendons at points of great pressure eg the patella
	Shin splints	Pain in the front of the tibia, usually due to fatigue fracture
	Shortening	Loss of length in a long bone
	Siderosis	A pneumoconiosis due to iron oxides
	Sign	What a doctor sees (compared with symptom, what is reported)
	Silicosis	A form of lung disease or pneumoconiosis caused by inhalation of silicon dust
	Simple fracture	An old term for fracture with intact skin
	Spastic	Excessive muscle tone
	Splint	Any device to prevent movement at a joint or fracture
	Spondylitis	Inflammation of the spine
	Spondylo-	(Prefix) Relating to a vertebra
	Spondylolisthesis	The slipping of one vertebra onto another
	Spondylolysis	A defect in a vertebra, usually due to fatigue fracture
	Spondylosis	Osteoarthritis of spine
	Sprain	A partial tear of a ligament
	Stenosis	Narrowing of an opening or passage in abnormal circumstances

	Stockholm Workshop Scale	A revised Taylor & Palmear Scale for the assessment of both Raynaud's phenomenon and sensorineural impairment in hand-arm vibration syndrome)
	Subluxation	Partial dislocation of a joint, so that the bone ends are mal-aligned but still in contact
	Sudek's atrophy	A variant of reflex sympathetic dystrophy which also involves wasting of muscle and thinning of bones (osteoporosis)
	Supination	Movement of forearm, so that the palm faces upwards and the thumb lies laterally
	Suture	A surgical stitch
	Symphysis	A joint with a disc of fibro-cartilage at the center and no cavity, which minimises movement
	Symptom	What the patient reports (compare with "sign")
	Syndesmosis	A fibrous joint in which the opposing surfaces are held together by a ligament
	Synovial effusion	Excess fluid generated in a synovial cavity near a joint
	Synovitis	Inflammation of membrane lining joint
T	Tachycardia	Rapid heart rate
	Tachypnoea	Rapid breathing
	Talcosis	A pneumoconiosis due to magnesium silicate
	Talipes	Club foot
	Tarsalgia	Aching pain in the sole of the foot
	Tarsus	Bones of foot between metatarsals and ankle
	Tendinitis	Inflammation of a tendon
	Taylor & Pelmear Scale	A scale for assessing the stages of Raynaud's phenomenon in HAVS (see Stockholm Scale)
	Temporary Threshold Shift	Reversible temporary increase in hearing threshold in response to loud noise
	Tenoplasty	Operation to repair a damaged tendon
	Tenorrhaphy	An operation to unite the ends of a severed tendon

Tenosynovitis	Inflammation of the sheath through which the tendon moves, usually extensor tendons of fingers at wrist level. Produces pain, difficulty in use, swelling and sometime an audible creaking (crepitus) on movement
Tensor	A muscle that stretches or tightens some part of the body
Tetraplegia	See Quadriplegia
Thenar	Relating to the ball of the thumb
Thoracic	Of the chest
Threshold of hearing	The minimum level of sound just audible
Threshold shift	The difference in dB between the hearing threshold levels measured on two occasions
Thrombosis	Formation of blood-clot within vessels or heart
Tinel's test	A test for carpal tunnel syndrome (percussion of median nerve to provoke symptoms)
Tinnitus	Noise heard in ear without external cause
Trabeculae	Bands of tissue which pass from the outer covering of an organ to the interior
Tracheostomy	Operation to open wind-pipe so that air may obtain direct entrance into lower air passages
Traction	Drawing or pulling of limb or spine as part of treatment
Trauma	Fracture or blow; overwhelmingly stressful event; often used in psychological sense
Trendelenburg test	Test for function of hip muscles
Trigger digit	Painful catching, triggering or locking of the flexor tendons of fingers or thumb
Triquetrum	Bone in wrist
Tuberosity	A rounded protuberance on the surface or side of a bone
Tympanic cavity	Middle ear
Tympanic membrane	Ear drum

U	Ulcer	Breach on surface of skin or membrane, which does not tend to heal quickly
	Ulna	The inner bone of the forearm
	Ulnar	Towards the same side of the forearm as the ulna
	Ulnar nerve	One of the principal nerves in the arm
	Ungual	Of the nails
	Ureter	Tubes which take urine from the pelvis of the kidney to the bladder
V	Valgus	A limb deformity in which the extremity moves away from the midline. The point of reference is proximal eg genu-valgum is knock knees
	Valvotomy	Surgical opening of a valve
	Varus	The opposite of valgus, eg bowed legs
	Vasospasm	The condition of arteries going into spasm
	Ventral	The side of the forearm on the same side as the palm
	Volar	On the same side of the hand as the palm (see palmar)
	Visual acuity	Sharpness of the vision
W	Watershed infarction	A stroke affecting the boundary between the parietal and occipital lobes of the brain on both sides
	Writer's Cramp	Pain, muscle spasm and tremor in the hand or forearm due to repetitive motions of fingers, hands or arms
X	Xerosis	Abnormal dryness, especially of eyes
	Xerostomia	Abnormal dryness of mouth

APPENDIX 3
PRESCRIBED DISEASES

List of Diseases covered by Industrial Injuries Disablement Benefit

List of diseases which are covered and the kinds of jobs which are included

The information in italics is an attempt to make the medical terms easier to understand and to give you some idea as to the jobs in which you might get one of the diseases. It is not a complete list of jobs and you should not be put off claiming just because your job is not listed. If in doubt, claim.

Disease Number	**Name of Disease or Injury Conditions due to physical agents (physical cause)**	**Type of job Any Job involving**
A1	Leukemia (other than chronic lymphatic leukemia) or cancer of the bone, female breast, testis or thyroid.	Exposure to electromagnetic radiations (other than radiant heat) or to ionising particles where the dose is sufficient to double the risk of occurrence of the condition. *For example, people working in the nuclear industry and hospital X-ray departments.*
A2	Cataract.	Frequent or prolonged exposure to radiation from red-hot or white-hot material. *For example, glass and metal workers, stokers.*
A3	Dysbarism, including decompression sickness, barotrauma and osteonecrosis. *For example, the bends.*	Subjection to compressed or rarefied air or other respirable gases or gaseous mixtures. *For example, underwater or tunnel workers.*

Disease Number	**Name of Disease or Injury Conditions due to physical agents (physical cause)**	**Type of job Any Job involving**
A4	Cramps of the hand or forearm due to repetitive movements. *For example, writer's cramp.*	Prolonged periods of handwriting, typing or other repetitive movements of the fingers, hand or arm. *For example, typists, clerks and routine assemblers.*
A5	Subcutaneous cellulitis of the hand. (Beat hand).	Manual labour causing severe or prolonged friction or pressure on the hand. *For example, miners and road workers using picks and shovels.*
A6	Bursitis or subcutaneous cellulitis arising at or about the knee due to severe or prolonged external friction or pressure at or about the knee. (Beat knee). *For example, housemaid's knee.*	Manual labour causing severe or prolonged external friction or pressure at or about the knee. *For example, workers who kneel a lot.*
A7	Bursitis or subcutaneous cellulitis arising at or about the elbow due to severe or prolonged external friction or pressure at or about the elbow. (Beat elbow).	Manual labour causing severe or prolonged external friction or pressure at or about the elbow. *For example, jobs involving continuous rubbing or pressure on the elbow.*
A8	Traumatic inflammation of the tendons of the hand or forearm or of the associated tendon sheaths. *Tenosynovitis.*	Manual labour, or frequent or repeated movements of the hand or wrist. *For example, routine assembly workers.*
A9	Miner's nystagmus. *Jerky movements of the eyeballs.*	Work in or about a mine.
A10	Occupational deafness. Sensorineural hearing loss amounting to at least 50dB in each ear, being the average of hearing losses at 1, 2 and 3KHz frequencies, and being due in the case of at least one ear to occupational noise.	(a) The use of powered (but not hand powered) grinding tools on metal (other than sheet metal or plate metal), or work wholly or mainly in the immediate vicinity of those tools whilst they are being so used; or (b) the use of pneumatic percussive tools on metal, or work wholly or mainly in the immediate vicinity of those tools whilst they are being so use, or

Disease Number	Name of Disease or Injury Conditions due to physical agents (physical cause)	Type of job Any Job involving
A10 *continued*		(c) the use of pneumatic percussive tools for drilling rock in quarries or underground or in mining coal or in sinking shafts for tunnelling in civil engineering works, or work wholly or mainly in the immediate vicinity of those tools whilst they are being so used; or (ca) the use of pneumatic percussive tools on stone in quarry works, or work wholly or mainly in the immediate vicinity of those tools whilst they are being so use; or (d) work wholly or mainly in the immediate vicinity of plant (excluding power press plant) engaged in the forging (including drop stamping) of metal by means of closed or open dies or drop hammers; or (e) work in textile manufacturing where the work is undertaken wholly or mainly in rooms or sheds in which there are machines engaged in weaving man-made or natural (including mineral) fibres or in the high speed false twisting of fibres; or (f) the use of, or work wholly or mainly in the immediate vicinity of machines engaged in cutting, shaping or cleaning metal nails; or (g) the use of, or work wholly or mainly in the immediate vicinity of, plasma spray guns engaged in the deposition of metal (h) the use of, or work wholly or mainly in the immediate vicinity of, any of the following machines engaged in the working of wood, that

Disease Number	Name of Disease or Injury Conditions due to physical agents (physical cause)	Type of job Any Job involving
A10 *continued*		is to say: multicutter moulding machines, planing machines, automatic or semi-automatic lathes, multiple cross-cut machines, automatic shaping machines, double end tenoning machines, vertical spindle moulding machines (including high speed routing machines), edge banding machines, saw banding machines with a blade width of not less than 75 millimetres and circular sawing machines in the operation of which the blade is moved towards the material being cut; or (i) the use of chain saws in forestry; or (j) air arc gouging or work wholly in or mainly in the immediate vicinity of air arc gouging; or (k) the use of band saws, circular saws or cutting disks for cutting metal in the metal founding or forging industries, or work wholly or mainly in the immediate vicinity of those tools whilst they are being so used; or (l) the use of circular saws for cutting products in the manufacture of steel, or work wholly or mainly in the immediate vicinity of those tools whilst they are being so used; or (m) the use of burners or torches for cutting or dressing steel based products, or work wholly or mainly in the immediate vicinity of those tools whilst they are being so used; or (n) work wholly or mainly in the immediate vicinity of skid transfer banks; or

Disease Number	Name of Disease or Injury Conditions due to physical agents (physical cause)	Type of job Any Job involving
A10 *continued*		(o) work wholly or mainly in the immediate vicinity of knock out and shake out grids in foundries; or (p) mechanical bobbin cleaning or work wholly or mainly in the immediate vicinity of mechanical bobbin cleaning; (q) the use of, or work wholly or mainly in the immediate vicinity of, vibrating metal moulding boxes in the concrete products industry; or (r) the use of, or work wholly or mainly in the immediate vicinity of, high pressure jets or water or a mixture of water and abrasive material in the water jetting industry (including work under water); or (s) work in ships' engine room; or (t) the use of circular saws for cutting concrete masonry blocks during manufacture, or work wholly or mainly in the immediate vicinity of those tools whilst they are being so used; or (u) burning stone in quarries by jet channelling processes, or work wholly or mainly in the immediate vicinity of such processes; or (v) work on gas turbines in connection with: (i) Performance testing on test bed; (ii) Installation testing of replacement engines in aircraft; (iii) Acceptance testing of Armed Service fixed wing combat planes; or (w) the use of, or work wholly or mainly in the immediate vicinity of: (i) Machines for automatic moulding, automatic

Disease Number	Name of Disease or Injury Conditions due to physical agents (physical cause)	Type of job Any Job involving
A10 *continued*		blow moulding or automatic glass pressing and forming machines used in the manufacture of glass containers or hollow ware; (ii) spinning machines using compressed air to produce glass wool or mineral wool; (iii) continuous glass toughening furnaces.
A11	Episodic blanching, occurring throughout the year, affecting the middle or proximal phalanges, or in the case of a thumb the proximal phalanx, of: (a) in the case of a person with 5 fingers (including thumb) on one hand, any 3 of those fingers; or (b) in the case of a person with only 4 such fingers, any 2 of those fingers; or (c) in the case of a person with less than 4 such fingers, any one of those fingers or, as the case may be, the one remaining finger. (Vibration white finger).	(a) the use of hand-held chain saws in forestry; or (b) the use of hand-held rotary tools in grinding or in the sanding or polishing of metal, or the holding of material being ground, or metal being sanded or polished, by rotary tools; or (c) the use of hand-held percussive metalworking tools, or the holding of metal being worked upon by percussive tools, in riviting, caulking, chipping, hammering, fettling or swaging; or (d) the use of hand-held powered percussive drills or hand-held powered percussive hammers in mining, quarrying, demolition, or on roads or footpaths, including road construction; or (e) the holding of material being worked upon by pounding machines in shoe manufacture.
A12	Carpal tunnel syndrome.	The use of hand-held powered tools whose internal parts vibrate so as to transmit that vibration to the hand, but excluding those which are solely powered by hand.

Disease Number	Name of Disease or Injury Conditions due to physical agents (physical cause)	Type of job Any Job involving
B1	Anthrax.	Contact with animals infected with anthrax or the handling (including the loading or unloading or transport) of animal products or residues. *For example, glue and shaving brush makers.*
B2	Glanders.	Contact with equine animals or their carcasses. *For example, farm and slaughterhouse workers, and grooms handling horses.*
B3	Infection by leptospira. *For example, swamp fever, swineherd's disease, and Weil's disease.*	(a) Work in places which are, or are liable to be, infested by rats, field mice or voles, or other small mammals; or (b) work at dog kennels or the care or handling of dogs; or (c) contact with bovine animals or their meat products or pigs or their meat products. *For example, farm, veterinary, sewerage and slaughterhouse workers.*
B4	Ancylostomiasis. *Hookworm disease, rarely found in this country.*	Work in or about a mine.
B5	Tuberculosis. *TB infection.*	Contact with a source of tuberculous infection. *For example, doctors, nurses, ambulance crews, pathology technicians and social workers.*
B6	Extrinsic allergic alveolitis (including farmer's lung)	Exposure to moulds or fungal spores or heterologous proteins by reason of employment in: (a) agriculture, horticulture, forestry, cultivation of edible fungi or maltworking; or (b) loading or unloading or handling in storage mouldy vegetable matter or edible fungi; or (c) caring for or handling birds; or (d) handling bagasse.

Disease Number	**Name of Disease or Injury Conditions due to physical agents (physical cause)**	**Type of job Any Job involving**
B7	Infection by organisms of the genus brucella. *Brucellosis.*	Contact with: (a) animals infected by brucella, or their carcasses or parts thereof, or their untreated products; or (b) laboratory specimens or vaccines of, or containing, brucella. *For example, farm, veterinary, slaughterhouse, animal laboratory workers.*
B8	Viral hepatitis. *An infection of the liver by a virus.*	Contact with: (a) human blood or human blood products; or (b) a source of viral hepatitis. *For example, doctors, nurses, ambulance crews, pathology workers.*
B9	Infection by Streptococcus suis. *A very rare form of meningitis from exposure to injected pigs or pork products.*	Contact with pigs infected by Streptococcus suis, or with the carcasses, products or residues of pigs so infected. *For example, pork butchers, pig breeders, slaughterhouse workers.*
B10(a)	Avian chlamydiosis.	Contact with birds infected with chlamydia psittaci, or with the remains or untreated products of such birds. *For example, duck farm workers, feather processing workers, poultry meat inspectors, pet shop owners and assistants.*
B10(b)	Ovine chlamydiosis.	Contact with sheep infected with chlamydia psittaci, or with the remains or untreated products of such sheep. *For example, sheep farm workers, veterinary surgeons.*
B11	Q fever.	Contact with animals, their remains or their untreated products. *For example, farm workers involved in the rearing of sheep, abattoir workers, veterinary surgeons.*

Disease Number	**Name of Disease or Injury Conditions due to physical agents (physical cause)**	**Type of job Any Job involving**
B12	Orf.	Contact with sheep or goats, or with the carcasses of sheep or goats. *For example, farm workers, abattoir workers, meat inspectors.*
B13	Hydatidosis.	Contact with dogs. *For example, shepherds, veterinarians and people who care for dogs.*
C1	Poisoning by lead or a compound of lead.	The use or handling of, and exposure to the fumes, dust or vapour of, lead or a compound of lead, or a substance containing lead. *For example, plumbers, painters, enamellers, pottery glazing workers.*
C2	Poisoning by manganese or a compound of manganese.	The use or handling of, or exposure to the fumes, dust or vapour of, manganese or a compound of manganese, or a substance containing manganese. *For example, dry battery, pottery glazing and soap workers.*
C3	Poisoning by phosphorus or an inorganic compound of phosphorus or poisoning due to the anticholinesterase or pseudo anticholinesterase action of organic phosphorus compounds.	The use or handling of, or exposure to the fumes, dust or vapour of, phosphorus or a compound of phosphorus, or a substance containing phosphorus. *For example, pest control, agricultural workers, workers on incendiary devices, match makers.*
C4	Poisoning by arsenic or a compound of arsenic.	The use or handling of, or exposure to the fumes, dust or vapour of, arsenic or a compound of arsenic, or a substance containing arsenic. *For example, leather, agricultural and metal pickling workers.*

Disease Number	Name of Disease or Injury Conditions due to physical agents (physical cause)	Type of job Any Job involving
C5	Poisoning by mercury or a compound of mercury.	The use or handling of, or exposure to the fumes, dust or vapour of, mercury or a compound of mercury, or a substance containing mercury. *For example, mirror/thermometer makers, market gardeners and explosive workers.*
C6	Poisoning by carbon bisulphide.	The use or handling of, or exposure to the fumes or vapour of, carbon bisulphide or a compound of carbon bisulphide, or a substance containing carbon bisulphide. *For example, artificial silk and cellophane makers.*
C7	Poisoning by benzene or a homologue of benzene. *Benzol/benzole, toluene/toluol, xylene/xylol.*	The use or handling of, or exposure to the fumes of, or vapour containing, benzene or any of its homologues. *For example, paint, dye, rubber goods and artificial leather workers.*
C8	Poisoning by nitro- or amino- or chloroderivative of benzene, or of a homologue of benzene, or poisoning by nitrochlorbenzene. *For example, Tri-nitrotoluene (TNT).*	The use or handling of, or exposure to the fumes of, or vapour containing, a nitro or amino- or chloro-derivative of benzene; or of a homologue of benzene, or nitrochlorbenzene. *For example, dyeing and chemical workers, solvents, disinfectants and wood preservation makers and users.*
C9	Poisoning by dinitrophenol or a homologue of dinitrophenol or by substituted dinitrophenols or by the salts of such substances. *Di-nitro-ortho-cresol (DNOC).*	The use or handling of, or exposure to the fumes of, or vapour containing, dinitrophenol or a homologue or substituted dinitrophenols or the salts of such substances. *For example, dye and wood preservative makers and users, agricultural workers.*
C10	Poisoning by tetrachloroethane.	The use or handling of, or exposure to the fumes of, or vapour containing, tetrachloroethane. *For example, photographic film, wax polish, adhesives, safety glass workers.*

Disease Number	**Name of Disease or Injury Conditions due to physical agents (physical cause)**	**Type of job Any Job involving**
C11	Poisoning by diethylene dioxide (dioxan). *Not dioxin (2 4 5 T).*	The use or handling of, or exposure to the fumes of, or vapour containing, diethylene dioxide (dioxan). *For example, polishing compounds, cosmetics and paint stripper makers.*
C12	Poisoning by methyl bromide	The use or handling of, or exposure to the fumes of, or vapour containing, methyl bromide. *For example, pest controllers, makers and users of fire extinguishers.*
C13	Poisoning by chlorinated naphthalene.	The use or handling of, or exposure to the fumes of, or dust or vapour containing, chlorinated naphthalene. *For example, synthetic wax and insulated wire makers.*
C14	Poisoning by nickel carbonyl.	Exposure to nickel carbonyl gas. *For example, nickel refinery workers.*
C15	Poisoning by oxides of nitrogen.	Exposure to oxides of nitrogen. *For example, explosives and nitric acid workers.*
C16	Poisoning by gonioma kamassi. *(African boxwood).*	The manipulation of gonioma kamassi or any process in or incidental to the manufacture of articles therefrom. *For example, weaving shuttle makers.*
C17	Poisoning by beryllium or a compound or beryllium.	The use or handling of, or exposure to the fumes, dust or vapour of, beryllium or a compound of beryllium, or a substance containing beryllium. *Beryllium (or glucinum) is found in the manufacture of fluorescent lights, neon signs, metallic alloys, atomic energy, radio valves, crucibles and electrical porcelain.*
C18	Poisoning by cadmium.	Exposure to cadmium dust or fumes. *For example, alkaline battery, jewellery and fluorescent light makers. Nuclear reactor workers.*

Disease Number	**Name of Disease or Injury Conditions due to physical agents (physical cause)**	**Type of job Any Job involving**
C19	Poisoning by acrylamide monomer.	The use or handling of, or exposure to, acrylamide monomer. *For example, paper, adhesive, dye, artificial leather, photographic emulsion makers.*
C20	Dystrophy of the cornea (including ulceration of the corneal surface) of the eye. *Wasting and ulceration of the corneal surface of the eye.*	(a) The use of handling of, or exposure to, arsenic, tar, pitch, bitumen, mineral oil (including paraffin), soot or any compound, product or residue of any of these substances, except quinone or hydroquinone; or (b) exposure to quinone or hydroquinone during their manufacture. *For example, chemical workers.*
C21(a) C21(b)	Localised new growth of the skin, papillomatous or keratotic. *Warts and scaliness.* Squamous-celled carcinoma of the skin. *A form of skin cancer/chimney sweep's cancer.*	The use or handling of, exposure to, arsenic, tar, pitch, bitumen, mineral oil (including paraffin), soot or any compound, product or residue of any of these substances, except quinone or hydroquinone. *For example, bituminous shale workers, optical lens makers, cotton mule spinners, workers exposed to tarry fumes.*
C22(a) C22(b)	Carcinoma of the mucous membrane of the nose or associated air sinuses. *Cancer of the lining of the nose or air sinuses.* Primary carcinoma of a bronchus or of a lung. *Cancer of the lung or bronchus.*	Work in a factory where nickel is produced by decomposition of a gaseous nickel compound which necessitates working in or about a building or buildings where that process or any other industrial process ancillary or incidental thereto is carried on.
C23	Primary neoplasm (including papilloma, carcinoma-in-situ and invasive carcinoma) of the epithelial lining of the urinary tract (renal pelvis, ureter, bladder and urethra). *Includes a form of cancer of the lining of the bladder or urinary tract.*	(a) Work in a building in which any of the following substances is produced for commercial purposes: (i) alpha-naphthylamine, betanaphthylamine or methylene-bisorthochloroaniline;

Disease Number	Name of Disease or Injury Conditions due to physical agents (physical cause)	Type of job Any Job involving
C23 *continued*		(ii) diphenyl substituted by at least one nitro or primary amino group or by at least one nitro and primary amino group (including benzidine); (iii) any of the substances mentioned in sub-paragraph (ii) above if further ring substituted by halogeno, methyl or methoxy groups, but not by other groups; (iv) the salts of any of the substances mentioned in the subparagraphs (i) to (iii) above; (v) auramine or magenta; or (b) the use or handling of any of the substances mentioned in sub-paragraph (a)(i) to (iv), or work in a process in which any such substance is used, handled or liberated; or (c) the maintenance or cleaning of any plant or machinery used in any such process as is mentioned in sub-paragraph (b), or the cleaning of clothing used in any such building as is mentioned in subparagraph (a) if such clothing is cleaned within the works of which the building forms a part or in a laundry maintained and used solely in connection with such works; (d) exposure to coal tar pitch volatiles produced in aluminium smelting involving the Soderberg process (the method of producing aluminium by electrolysis in which the anode consists of a paste of petroleum coke and mineral oil which is baked in situ).

Disease Number	Name of Disease or Injury Conditions due to physical agents (physical cause)	Type of job Any Job involving
C23 *continued*		*For example, gas retort workers, laboratory workers, workers in the synthetic dye, rubber, cable and chemical industries.*
C24(a) C24(b) C24(c)	Angiosarcoma of the liver. *A form of liver cancer.* Osteolysis of the terminal phalanges of the fingers. *A condition of the bones of the fingertips.* Non-cirrhotic portal fibrosis. *A form of liver damage.*	(a) Work in or about machinery or apparatus used for the polymerization of vinyl chloride monomer, a process which, for the purposes of this provision, comprises all operations up to and including the drying of the slurry produced by the polymerization and the packaging of the dried product; or (b) work in a building or structure in which any part of that process takes place. *For example, PVC makers.*
C25	Occupational vitiligo. *White patches on the skin.*	The use or handling of, or exposure to, para-tertiary-butylphenol, para-tertiarybutylcatechol, para-amyl-phenol, hydroquinone or the monobenzyl or monobutyl ether of hydroquinone. *For example, car, shoe or chemical workers.*
C26	Damage to the liver or kidneys due to exposure to Carbon Tetrachloride.	The use of or handling of, or exposure to the fumes of, or vapour containing Carbon Tetrachloride. *For example, workers in chemical factories.*
C27	Damage to the liver or kidneys due to exposure to Carbon Trichloromethane (Chloroform).	The use of or handling of, or exposure to the fumes of, or vapour containing Trichloromethane (Chloroform). *For example, workers in chemical factories, laboratory workers.*
C28	Central nervous system dysfunction and associated gastro-intestinal disorders due to exposure to Chloromethane (Methyl Chloride).	The use of or handling of, or exposure to the fumes of, or vapour containing Chloromethane (Methyl Chloride). *For example, workers in chemical factories, laboratory workers.*

Disease Number	Name of Disease or Injury Conditions due to physical agents (physical cause)	Type of job Any Job involving
C29	Peripheral neuropathy due to exposure to n-hexane or methyl n-butyl ketone.	The use of or handling of, or exposure to the fumes of, or vapour containing nhexane or methyl n-butyl ketone. *For example, printers, workers exposed to fuel emissions or adhesive fumes.*
C30	Chrome dermatitis, or ulceration of the mucous membranes or the epidermis, resulting from exposure to chromic acid, chromates or bi-chromates. *For example, ulcers of the inside of the nose or mouth, or skin rash, dermatitis through working with chromium compounds.*	The use or handling of, or exposure to, chromic acid, chromates or bi-chromates. *For example, workers in chemical factories, dye factories, photographic processors, leather tanning industry, chromium plating industry.*
D1	Pneumoconiosis. *Includes silicosis and asbestosis.*	(1) (a) The mining, quarrying or working of silica rock or the working of dried quartzose sand or any dry deposit or dry residue of silica or any dry admixture containing such materials (including any occupation in which any of the aforesaid operations are carried out incidentally to the mining or quarrying of other minerals or to the manufacture of articles containing crushed or ground silica rock); (b) the handling of any of the materials specified in the foregoing subparagraph in or incidental to any of the operations mentioned therein, or substantial exposure to the dust arising from such operations. (2) The breaking, crushing or grinding of flint or the working or handling of broken, crushed or ground

Disease Number	Name of Disease or Injury Conditions due to physical agents (physical cause)	Type of job Any Job involving
D1 *continued*		flint or materials containing such flint, or substantial exposure to the dust arising from any such operations. (3) Sand blasting by means of compressed air with the use of quartzone sand or crushed silica rock or flint, or substantial exposure to the dust arising from sand and blasting. (4) Work in a foundry or the performance of, or substantial exposure to the dust arising from, any of the following operations: (a) the freeing of steel castings from adherent siliceous substance; (b) the freeing of metal castings, from adherent siliceous substance: (i) by blasting with an abrasive propelled by compressed air, by steam or by a wheel, or (ii) by the use of power-driven tools (5) The manufacture of china or earthenware (including sanitary earthenware, electrical earthenware and earthenware tiles), and any occupation involving substantial exposure to the dust arising therefrom. (6) The grinding of mineral graphite, or substantial exposure to the dust arising from such grinding. (7) The dressing of granite or any igneous rock by masons or the crushing of such materials, or substantial exposure to the dust arising from such operations. (8) The use, or preparation for use, of a grindstone, or substantial exposure to the dust arising therefrom.

Disease Number	Name of Disease or Injury Conditions due to physical agents (physical cause)	Type of job Any Job involving
D1 *continued*		(9) (a) The working or handling of asbestos or any admixture of asbestos; (b) the manufacture or repair of asbestos textiles or other articles, containing or composed of asbestos; (c) the cleaning of any machinery or plant used in any foregoing operations and of any chambers, fixtures and appliances for the collection of asbestos dust; (d) substantial exposure to the dust arising from any of the foregoing operations. (10) (a) Work underground in any mine in which one of the objects of the mining operations is the getting of any mineral; (b) the working or handling above ground at any coal or tin mine of any minerals extracted therefrom, or any operation incidental thereto; (c) the trimming of coal in any ship, barge, or lighter, or in any dock or harbour or at any wharf or quay; (d) the sawing, splitting or dressing of slate, or any operation incidental thereto. (11) The manufacture of carbon electrodes by an industrial undertaking for use in the electrolytic extraction of aluminium from aluminium oxide, and any occupation involving substantial exposure to the dust arising therefrom.

Disease Number	Name of Disease or Injury Conditions due to physical agents (physical cause)	Type of job Any Job involving
D1 *continued*		(12) Boiler scaling or substantial exposure to the dust arising therefrom. (13) Exposure to dust if the person employed in it has never at any time worked in any of the other occupations listed.
D2	Byssinosis. *A respiratory condition.*	Work in any room where any process up to and including the weaving process is performed in a factory in which the spinning or manipulation of raw or waste cotton or of flax, or the weaving of cotton or flax, is carried on. *For example, cotton or flax workers.*
D3	Diffuse mesothelioma (primary neoplasm of the mesothelium of the pleura or of the pericardium or of the peritoneum). *A cancer starting in the covering of the lungs or the lining of the abdomen.*	Exposure to asbestos, asbestos dust or any admixture of asbestos at a level above that commonly found in the environment at large.
D4	Allergic rhinitis which is due to exposure to any of the following agents: (a) isocyanates (b) platinum salts (c) fumes of dusts arising from the manufacture, transport or use of hardening agents (including epoxy resin curing agents) based on phthalic anhydride, tetrachlorophthalic anhydride, trimellitic anhydride or triethylenetetramine (d) fumes arising from the use of rosin as a soldering flux (e) proteolytic enzymes (f) animals including insects and other anthropods used for the purposes of research or education or in laboratories.	Exposure to any of the agents set out in column 2 of this paragraph. *Wide range of occupations for example, metal plating industry, food processing, laboratory workers, grain processing, drug manufacture, washing powder manufacture, hair dressing, electronic industry, welders, dye tea and coffee processing.*

Disease Number	**Name of Disease or Injury Conditions due to physical agents (physical cause)**	**Type of job Any Job involving**
D4 *continued*	(g) dusts arising from the sowing, cultivation, harvesting, drying, handling, milling, transport or storage of barley, oats, rye, wheat or maize, or the handling milling, transport or storage of meal or flour made therefrom (h) antibiotics (i) cimetidine (j) wood dust (k) ispaghula (l) castor bean dust (m) ipecacuanha (n) azodicarbonamide (o) animals including insects and other arthropods or their larval forms used for the purposes of pest control or fruit cultivation, or the larval forms of animals used for the purposes of research, education or in laboratories (p) glutaraldehyde (q) persulphate salts or henna (r) crustaceans or fish or products arising from these in the food processing industry (s) reactive dyes (t) soya bean (u) tea dust (v) green coffee bean dust (w) fumes from stainless steel welding. *For example, hay fever symptoms.*	
D5	Non-infective dermatitis of external origin (excluding dermatitis due to ionising particles or electromagnetic radiations other than radiant heat). *For example, skin rash, dermatitis.*	Exposure to dust, liquid or vapour or any other external agent except chromic acid, chromates or bi-chromates, capable of irritating the skin (including friction or heat but excluding ionising particles or electromagnetic radiations other than radiant heat). *For example, any job involving exposure to a substance which can irritate the skin except for jobs involving exposure to chromium compounds (see C30) and radiation.*

Disease Number	Name of Disease or Injury Conditions due to physical agents (physical cause)	Type of job Any Job involving
D6	Carcinoma of the nasal cavity or associated air sinuses (nasal carcinoma). *Cancer of the nose.*	(a) Attendance for work in or about a building where wooden goods are manufactured or repaired; or (b) attendance for work in a building used for the manufacture of footwear or components of footwear made wholly or partly of leather or fibreboard; or (c) attendance for work at a place used wholly or mainly for the repair of footwear made wholly or partly of leather or fibreboard.
D7	Asthma which is due to exposure to any of the following agents: (a) isocyanates (b) platinum salts (c) fumes or dusts arising from the manufacture, transport or use of hardening agents (including epoxy resin curing agents) based on phthalic anhydride, tetrachlorophthalic anhydride, trimellitic anhydride or triethylenetetramine (d) fumes arising from the use of rosin as a soldering flux (e) proteolytric enzymes (f) animals including insects and other arthropods used for the purposes of research or education or in laboratories (g) dusts arising from the sowing, cultivation, harvesting, drying, handling, milling, transport or storage of barley, oats, rye, wheat or maize, or the handling, milling, transport or storage of meal for flour made therefrom (h) antibiotics (i) cimetidine (j) wood dust (k) ispaghula	Exposure to any of the agents set out in column 2 of this paragraph.

Disease Number	**Name of Disease or Injury Conditions due to physical agents (physical cause)**	**Type of job Any Job involving**
D7 *continued*	(l) castor bean dust (m) ipecacuanha (n) azodicarbonamide (o) animals including insects and other arthropods or their larval forms, used for the purposes of pest control or fruit cultivation, or the larval forms of animals used for the purposes of research, education or in laboratories (p) glutaraldehyde (q) persulphate salts or henna (r) crustaceans or fish or products arising from these in the food processing industry (s) reactive dyes (t) soya bean (u) tea dust (v) green coffee bean dust (w) fumes from stainless steel welding (x) any other sensitising agent (Occupational asthma)	
D8	Primary carcinoma of the lung where there is accompanying evidence of one or both of the following: (a) asbestosis (b) unilateral or bilateral diffuse pleural thickening extending to a thickness of 5mm or more at any point within the area affected as measured by a plain chest radiograph (not being a computerised tomography scan or other form of imaging) which: (i) in the case of unilateral diffuse pleural thickening, covers 50% or more of the area of the chest wall of the lung affected; or (ii) in the case of bilateral diffuse pleural thickening, covers 25% or more of the combined area of the chest wall of both lungs.	(a) The working or handling of asbestos or any admixture of asbestos; or (b) the manufacture or repair of asbestos textiles or other articles containing or composed of asbestos; or (c) the cleaning of any machinery or plant used in any of the foregoing operations and of any chambers, fixtures and appliances for the collection of asbestos dust; or (d) substantial exposure to the dust arising from any of the foregoing operations.

Disease Number	Name of Disease or Injury Conditions due to physical agents (physical cause)	Type of job Any Job involving
D9	Unilateral or bilateral diffuse pleural thickening extending to a thickness extending to a thickness of 5mm or more at any point within the area affected as measured by a plain chest radiograph (not being a computerised tomography scan or other form of imaging) which: (i) in the case of unilateral diffuse pleural thickening, covers 50 % or more of the area of the chest wall of the lung affected; or (ii) in the case of bilateral diffuse pleural thickening, covers 25% or more of the combined area of the chest wall of both lungs.	As D8 above.
D10	Primary carcinoma of the lung.	(a) Work underground in a tin mine; or (b) exposure to bis (chloromethyl) ether produced during the manufacture of chloromethyl methyl ether; or (c) exposure to zinc chromate, calcium chromate or strontium chromate in their pure forms.
D11	Primary carcinoma of the lung where there is accompanying silicosis.	Exposure to silica dust in the course of: (a) the manufacture of glass or pottery (b) tunnelling in, or quarrying sandstone or granite (c) mining metal ores (d) slate quarrying or the manufacturing of artefacts from slate (e) mining clay (f) using siliceous materials as abrasives (g) cutting stone (h) stonemasonry (i) work in a foundry.

Disease Number	Name of Disease or Injury Conditions due to physical agents (physical cause)	Type of job Any Job involving
D12	Chronic bronchitis or emphysema; or both where, with maximum effort, the is accompanying evidence of a forced expiratory volume in one second which is: (i) at least one litre below the appropriate mean value predicted, obtained from the following prediction formulae which gives the mean values predicted in litres: • For a man, where the measurement is made without back-extrapolation, (3.62 x Height in metres) minus (0.031 x Age in years) minus 1.41; or, where the measurement is made with back-extrapolation, (3.71 x Height in metres) minus (0.032 x Age in years) minus 1.44 • For a woman, where the measurement is made without back-extrapolation, (3.29 x Height in metres) minus (0.029 x Age in years) minus 1.42; or, where the measurement is made with back-extrapolation, (3.37 x Height in metres) minus (0.030 x Age in years) minus 1.46 (b) **or** less than one litre.	Exposure to coal dust by reason of working underground in a coal mine for a period or periods amounting in the aggregate to at least 20 years (whether before or after 5 July 1948) and any such period or periods shall include a period or periods of incapacity while engaged in such an occupation.

APPENDIX 4
CONTROL OF ASBESTOS AT WORK REGULATIONS 2002

STATUTORY INSTRUMENTS

2002 No. 2675

HEALTH AND SAFETY

Control of Asbestos at Work Regulations 2002

Made	*24th October 2002*
Laid before Parliament	*31st October 2002*
Coming into force	
All regulations except regulations 4 and 20	*21st November 2002*
Regulation 4	*21st May 2004*
Regulation 20	*21st November 2004*

ARRANGEMENT OF REGULATIONS

12. Maintenance of control measures etc.
13. Provision and cleaning of protective clothing.
14. Arrangements to deal with accidents, incidents and emergencies.
15. Duty to prevent or reduce the spread of asbestos.
16. Cleanliness of premises and plant.
17. Designated areas.
18. Air monitoring.
19. Standards for air testing.
20. Standards for analysis.
21. Health records and medical surveillance.
22. Washing and changing facilities.
23. Storage, distribution and labelling of raw asbestos and asbestos waste.
24. Supply of products containing asbestos for use at work.
25. Exemption certificates.
26. Extension outside Great Britain.
27. Revocations, amendments and savings.
28. Defence.

Schedule 1

Particulars to be included in a notification.

Schedule 2

The labelling of raw asbestos, asbestos waste and products containing asbestos.

The Secretary of State in the exercise of the powers conferred on him by sections 15(1), (2), (4), (5)(b), (6)(b) and (9), 18(2) and 82(3)(a) of, and paragraphs 1(1) to (4), 3(2), 6(1), 8 to 11, 14, 15(1), 16 and 20 of Schedule 3 to, the Health and Safety at Work etc. Act 1974[1] ("the 1974 Act ") and of all other powers enabling him in that behalf and for the purpose of giving effect without modifications to proposals submitted to him by the Health and Safety Commission under section 11(2)(d) of the 1974 Act after the carrying out by the said Commission of consultations in accordance with section 50(3) of that Act, hereby makes the following Regulations:

[1] 1974 c.37; sections 11(2), 15(1) and 50(3) were amended by the Employment Protection Act 1975 (c.71), Schedule 15, paragraphs 4, 6 and 16(3) respectively.

Citation and commencement

1. These Regulations may be cited as the Control of Asbestos at Work Regulations 2002 and shall come into force on 21st November 2002 except regulation 4 which shall come into force on 21st May 2004 and regulation 20 which shall come into force on 21st November 2004.

Interpretation

2.– (1) In these Regulations –

"action level" means one of the following cumulative exposures to asbestos over a continuous 12-week period when measured or calculated by a method approved by the Health and Safety Commission, namely –

(a) where the exposure is solely to chrysotile, 72 fibre-hours per millilitre of air;

(b) where exposure is to any other form of asbestos either alone or in mixtures including mixtures of chrysotile with any other form of asbestos, 48 fibre-hours per millilitre of air; or

(c) where both types of exposure occur separately during the 12-week period concerned, a proportionate number of fibre-hours per millilitre of air;

"adequate" means adequate having regard only to the nature and degree of exposure to asbestos and "adequately" shall be construed accordingly;

"appointed doctor" means a registered medical practitioner appointed for the time being in writing by the Executive for the purpose of these Regulations;

"approved" means approved for the time being in writing by the Health and Safety Commission or the Executive as the case may be;

"asbestos" means the following minerals, that is to say, crocidolite, amosite, chrysotile, fibrous actinolite, fibrous anthophyllite or fibrous tremolite or any mixture containing any of those minerals;

"asbestos area" and "respirator zone" shall be construed in accordance with regulation 17;

"control limit" means one of the following concentrations of asbestos in the atmosphere when measured or calculated by the reference method described in Annex I to Council Directive 83/477/EEC[2] or by a method giving equivalent

[2] OJ No. L 263, 24.9.83, p. 25.

results to that reference method approved by the Health and Safety Commission, namely –

(a) for chrysotile –

(i) 0.3 fibres per millilitre of air averaged over a continuous period of 4 hours,

(ii) 0.9 fibres per millilitre of air averaged over a continuous period of 10 minutes;

(b) for any other form of asbestos either alone or in mixtures including mixtures of chrysotile with any other form of asbestos –

(i) 0.2 fibres per millilitre of air averaged over a continuous period of 4 hours,

(ii) 0.6 fibres per millilitre of air averaged over a continuous period of 10 minutes;

"control measure" means a measure taken to prevent or reduce exposure to asbestos (including the provision of systems of work and supervision, the cleaning of workplaces, premises, plant and equipment, the provision and use of engineering controls and personal protective equipment);

"employment medical adviser" means an employment medical adviser appointed under section 56 of the Health and Safety at Work etc. Act 1974;

"ISO 17025" means European Standard EN ISO/IEC 17025, "General requirements for the competence of testing and calibration laboratories" accepted by the Comité Européen de Normalisation and the Comité Européen de Normalisation Electrotechnique (CEN/CENELEC) on 6th December 1999[3] as revised or reissued from time to time;

"medical examination" includes any laboratory tests and X-rays that a relevant doctor may require;

"personal protective equipment" means all equipment (including clothing) which is intended to be worn or held by a person at work and which protects that person against one or more risks to his health, and any addition or accessory designed to meet that objective;

"relevant doctor" means an appointed doctor or an employment medical adviser;

"the risk assessment" means the assessment of risk required by regulation 6(1)(a).

---

[3] Reference number EN ISO/IEC 17025: 2000.

(2) For the purposes of these Regulations, except as specified in regulation 10(4), in determining whether an employee is exposed to asbestos or whether the extent of such exposure exceeds the action level or a relevant control limit, no account shall be taken of respiratory protective equipment which, for the time being, is being worn by that employee.

Duties under these Regulations

3.– (1) Where a duty is placed by these Regulations on an employer in respect of his employees, he shall, so far as is reasonably practicable, be under a like duty in respect of any other person, whether at work or not, who may be affected by the work activity carried out by the employer except that the duties of the employer –

(a) under regulation 9 (information, instruction and training) shall not extend to persons who are not his employees unless those persons are on the premises where the work is being carried out; and

(b) under regulation 21 (health records and medical surveillance) shall not extend to persons who are not his employees.

(2) These Regulations shall apply to a self-employed person as they apply to an employer and an employee and as if that self-employed person were both an employer and an employee.

(3) Regulation 16(1), insofar as it requires an employer to ensure that premises are thoroughly cleaned, shall not apply to –

(a) a fire authority in respect of premises attended, for the purpose of fighting a fire or in an emergency, by a fire brigade maintained by that authority; or

(b) the employer of persons who attend a ship in dock premises for the purpose of fighting a fire or in an emergency, in respect of any ship so attended,

and in this regulation, "ship" includes all vessels and hovercraft which operate on water or land and water, and "dock premises" means a dock, wharf, quay, jetty or other place at which ships load or unload goods or embark or disembark passengers, together with neighbouring land or water which is used or occupied, or intended to be used or occupied, for those or incidental activities, and any part of a ship when used for those or incidental activities.

Duty to manage asbestos in non-domestic premises

4.– (1) In this regulation "the dutyholder" means –

(a) every person who has, by virtue of a contract or tenancy, an obligation of any extent in relation to the maintenance or repair of non-domestic premises or any means of access thereto or egress therefrom; or

(b) in relation to any part of non-domestic premises where there is no such contract or tenancy, every person who has, to any extent, control of that part of those non-domestic premises or any means of access thereto or egress therefrom,

and where there is more than one dutyholder, the relative contribution to be made by each such person in complying with the requirements of this regulation will be determined by the nature and extent of the maintenance and repair obligation owed by that person.

(2) Every person shall cooperate with the dutyholder so far as is necessary to enable the dutyholder to comply with his duties under this regulation.

(3) In order to enable him to manage the risk from asbestos in non-domestic premises, the dutyholder shall ensure that a suitable and sufficient assessment is carried out as to whether asbestos is or is liable to be present in the premises.

(4) In making the assessment –

(a) such steps as are reasonable in the circumstances shall be taken; and

(b) the condition of any asbestos which is, or has been assumed to be, present in the premises shall be considered.

(5) Without prejudice to the generality of paragraph (4), the dutyholder shall ensure that –

(a) account is taken of building plans or other relevant information and of the age of the premises; and

(b) an inspection is made of those parts of the premises which are reasonably accessible.

(6) The dutyholder shall ensure that the assessment is reviewed forthwith if –

(a) there is reason to suspect that the assessment is no longer valid; or

(b) there has been a significant change in the premises to which the assessment relates.

(7) The dutyholder shall ensure that the conclusions of the assessment and every review are recorded.

(8) Where the assessment shows that asbestos is or is liable to be present in any part of the premises the dutyholder shall ensure that –

(a) a determination of the risk from that asbestos is made;

(b) a written plan identifying those parts of the premises concerned is prepared; and

(c) the measures which are to be taken for managing the risk are specified in the written plan.

(9) The measures to be specified in the plan for managing the risk shall include adequate measures for –

(a) monitoring the condition of any asbestos or any substance containing or suspected of containing asbestos;

(b) ensuring any asbestos or any such substance is properly maintained or where necessary safely removed; and

(c) ensuring that information about the location and condition of any asbestos or any such substance is –

(i) provided to every person liable to disturb it, and

(ii) made available to the emergency services.

(10) The dutyholder shall ensure that –

(a) the plan is reviewed and revised at regular intervals, and forthwith if –

(i) there is reason to suspect that the plan is no longer valid, or

(ii) there has been a significant change in the premises to which the plan relates;

(b) the measures specified in the plan are implemented; and

(c) the measures taken to implement the plan are recorded.

(11) In this regulation, a reference to –

(a) "the assessment" is a reference to the assessment required by paragraph (3);

(b) "the premises" is a reference to the non-domestic premises referred to in paragraph (1); and

(c) "the plan" is a reference to the plan required by paragraph (8).

Identification of the type of asbestos

5. An employer shall not carry out work which exposes or is liable to expose his employees to asbestos unless either –

(a) before commencing that work, he has identified, by analysis or otherwise, the type of asbestos involved in the work; or

(b) he has assumed that the asbestos is not chrysotile alone and for the purposes of these Regulations has treated it accordingly.

Assessment of work which exposes employees to asbestos

6.– (1) An employer shall not carry out work which is liable to expose his employees to asbestos unless he has –

(a) made a suitable and sufficient assessment of the risk created by that exposure to the health of those employees and of the steps that need to be taken to meet the requirements of these Regulations;

(b) recorded the significant findings of that risk assessment as soon as is practicable after the risk assessment is made; and

(c) implemented the steps referred to in sub-paragraph (a).

(2) Without prejudice to the generality of paragraph (1), the risk assessment shall –

(a) subject to regulation 5, identify the type of asbestos to which employees are liable to be exposed;

(b) determine the nature and degree of exposure which may occur in the course of the work;

(c) consider the effect of control measures which have been or will be taken in accordance with regulation 10;

(d) consider the results of monitoring of exposure in accordance with regulation 18;

(e) set out the steps to be taken to prevent that exposure or reduce it to the lowest level reasonably practicable;

(f) consider the results of relevant medical surveillance; and

(g) include such additional information as the employer may need in order to complete the risk assessment.

(3) The risk assessment shall be reviewed regularly and forthwith if –

(a) there is reason to suspect that the existing risk assessment is no longer valid;

(b) there is a significant change in the work to which the risk assessment relates; or

(c) the results of any monitoring carried out pursuant to regulation 18 show it to be necessary,

and where, as a result of the review, changes to the risk assessment are required, those changes shall be made.

(4) Where, in accordance with the requirement in paragraph (2)(b), the risk assessment has determined that the exposure of his employees to asbestos may exceed the action level, the employer shall keep a copy of the significant findings of that risk assessment at those premises at which, and for such time as, the work to which that risk assessment relates is being carried out.

Plans of work

7.– (1) An employer shall not undertake any work with asbestos unless he has prepared a suitable written plan of work (hereinafter referred to in this regulation as "the plan of work") detailing how that work is to be carried out.

(2) The employer shall keep a copy of the plan of work –

(a) at those premises at which; and

(b) for such time as,

the work to which the plan relates is being carried out.

(3) The plan of work shall include in particular details of –

(a) the nature and probable duration of the work;

(b) the location of the place where the work is to be carried out;

(c) the methods to be applied where the work involves the handling of asbestos or materials containing asbestos;

(d) the characteristics of the equipment to be used for –

(i) protection and decontamination of those carrying out the work,

(ii) protection of other persons on or near the worksite; and

(e) the measures which the employer intends to take in order to comply with the requirements of regulation 10.

(4) The employer shall ensure, so far as is reasonably practicable, that the work to which the plan of work relates is carried out in accordance with that plan.

Notification of work with asbestos

8.– (1) This regulation shall apply to work at the premises of the employer in which an employee is or is liable to be exposed to asbestos unless –

(a) the extent of that exposure neither exceeds nor is liable to exceed the action level; or

(b) the work is work to which the Asbestos (Licensing) Regulations 1983[4] apply.

(2) An employer shall not carry out work to which this regulation applies for the first time unless he has notified the enforcing authority in writing of the particulars specified in Schedule 1 at least 14 days before commencing that work or before such shorter time as the enforcing authority may agree.

(3) Where an employer has notified work in accordance with paragraph (2) and there is a material change in that work which might affect the particulars so notified (including the cessation of the work), the employer shall forthwith notify the enforcing authority of that change.

Information, instruction and training

9.– (1) Every employer shall ensure that adequate information, instruction and training is given to those of his employees –

(a) who are or are liable to be exposed to asbestos, or who supervise such employees, so that they are aware of –

(i) the significant findings of the risk assessment,

(ii) the risks to health from asbestos,

(iii) the precautions which should be observed, and

(iv) the relevant control limit and action level,

[4] S.I. 1983/1649, as amended by S.I. 1998/3233.

in order to safeguard themselves and other employees; and

(b) who carry out work in connection with the employer's duties under these Regulations, so that they can carry out that work effectively.

(2) The information, instruction and training required by paragraph (1) shall be –

(a) given at regular intervals;

(b) adapted to take account of significant changes in the type of work carried out or methods of work used by the employer; and

(c) provided in a manner appropriate to the nature and degree of exposure identified by the risk assessment.

Prevention or reduction of exposure to asbestos

10.– (1) Every employer shall –

(a) prevent the exposure of his employees to asbestos so far as is reasonably practicable;

(b) where it is not reasonably practicable to prevent such exposure –

(i) reduce the exposure of his employees to asbestos to the lowest level reasonably practicable by measures other than the use of respiratory protective equipment, and

(ii) ensure that the number of his employees who are exposed to asbestos is as low as is reasonably practicable.

(2) Without prejudice to the generality of paragraph (1), where employees may be exposed to asbestos in a manufacturing process or in the installation of a product, prevention of such exposure to asbestos shall be achieved, where it is practicable, by substituting for asbestos a substance which, under the conditions of its use, does not create a risk to the health of his employees or creates a lesser risk than that created by asbestos.

(3) Where it is not practicable to avoid the use of asbestos by substitution in accordance with paragraph (2), the measures referred to in paragraph (1)(b)(i) shall include, in order of priority –

(a) the design and use of appropriate work processes, systems and engineering controls and the provision and use of suitable work equipment and materials in order to avoid or minimise the release of asbestos; and

(b) the control of exposure at source, including adequate ventilation systems and appropriate organisational measures.

(4) Where, in pursuance of paragraph (1), it is not reasonably practicable to reduce the exposure of an employee to below the control limits which apply to that exposure, then, in addition to taking the measures required by that paragraph, the employer shall provide that employee with suitable respiratory protective equipment which will reduce the concentration of asbestos in the air inhaled by the employee (after taking into account the effect of that respiratory protective equipment) to a concentration which is as low as is reasonably practicable and in any event below those control limits.

(5) Personal protective equipment provided by an employer in accordance with this regulation or regulation 13(1) shall be suitable for the purpose and shall –

(a) comply with any provision in the Personal Protective Equipment Regulations 2002[5] which is applicable to that item of personal protective equipment; or

(b) in the case of respiratory protective equipment, where no provision referred to in sub-paragraph (a) applies, be of a type approved or shall conform to a standard approved, in either case, by the Executive.

(6) Every employer shall take immediate steps to remedy the situation where the concentration of asbestos in air inhaled by any employee exceeds the relevant control limit.

(7) Where asbestos is used in a work process, or is produced by a work process, the employer shall ensure that the quantity of asbestos and materials containing asbestos at the premises where that work is carried out is reduced to as low a level as is reasonably practicable.

Use of control measures etc.

11.– (1) Every employer who provides any control measure, other thing or facility pursuant to these Regulations shall ensure so far as is reasonably practicable that it is properly used or applied as the case may be.

[5] S.I. 2002/1144.

(2) Every employee shall make full and proper use of any control measure, other thing or facility provided pursuant to these Regulations and, where relevant, shall –

(a) take all reasonable steps to ensure it is returned after use to any accommodation provided for it; and

(b) if he discovers a defect therein report it forthwith to his employer.

Maintenance of control measures etc.

12.– (1) Every employer who provides any control measure to meet the requirements of these Regulations shall ensure that, where relevant, it is maintained in an efficient state, in efficient working order, in good repair and in a clean condition.

(2) Where exhaust ventilation equipment or respiratory protective equipment (except disposable respiratory protective equipment) is provided to meet the requirements of these Regulations, the employer shall ensure that thorough examinations and tests of that equipment are carried out at suitable intervals by a competent person.

(3) Every employer shall keep a suitable record of the examinations and tests carried out in accordance with paragraph (2) and of repairs carried out as a result of those examinations and tests, and that record or a suitable summary thereof shall be kept available for at least five years from the date on which it was made.

Provision and cleaning of protective clothing

13.– (1) Every employer shall provide adequate and suitable protective clothing for such of his employees as are exposed or are liable to be exposed to asbestos, unless no significant quantity of asbestos is liable to be deposited on the clothes of the employee while he is at work.

(2) The employer shall ensure that protective clothing provided in pursuance of paragraph (1) is either disposed of as asbestos waste or adequately cleaned at suitable intervals.

(3) The cleaning required by paragraph (2) shall be carried out either on the premises where the exposure to asbestos has occurred, where those premises are suitably equipped for such cleaning, or in a suitably equipped laundry.

(4) The employer shall ensure that protective clothing which has been used and is to be removed from the premises referred to in paragraph (3) (whether for cleaning, further use or disposal) is packed, before being removed, in a suitable

container which shall be labelled in accordance with the provisions of Schedule 2 as if it were a product containing asbestos or, in the case of protective clothing intended for disposal as waste, in accordance with regulation 23(3).

(5) Where, as a result of the failure or improper use of the protective clothing provided in pursuance of paragraph (1), a significant quantity of asbestos is deposited on the personal clothing of an employee, then for the purposes of paragraphs (2), (3) and (4) that personal clothing shall be treated as if it were protective clothing.

Arrangements to deal with accidents, incidents and emergencies

14.– (1) Subject to paragraph (3) and without prejudice to the relevant provisions of the Management of Health and Safety at Work Regulations 1999[6], in order to protect the health of his employees from an accident, incident or emergency related to the use of asbestos in a work process or to the removal or repair of asbestos-containing materials at the workplace, the employer shall ensure that –

(a) procedures, including the provision of relevant safety drills (which shall be tested at regular intervals), have been prepared which can be put into effect when such an event occurs;

(b) information on emergency arrangements, including –

(i) details of relevant work hazards and hazard identification arrangements, and

(ii) specific hazards likely to arise at the time of an accident, incident or emergency,

is available; and

(c) suitable warning and other communication systems are established to enable an appropriate response, including remedial actions and rescue operations, to be made immediately when such an event occurs.

(2) The employer shall ensure that information on the procedure and systems required by paragraph (1)(a) and (c) and the information required by paragraph (1)(b) is –

(a) made available to relevant accident and emergency services to enable those services, whether internal or external to the workplace, to prepare their own response procedures and precautionary measures; and

[6] S.I. 1999/3242.

(b) displayed at the workplace, if this is appropriate.

(3) Paragraph (1) shall not apply where –

(a) the results of the risk assessment show that, because of the quantity of asbestos present at the workplace, there is only a slight risk to the health of employees; and

(b) the measures taken by the employer to comply with the duty under regulation 10(1) are sufficient to control that risk.

(4) In the event of an accident, incident or emergency related to the unplanned release of asbestos at the workplace, the employer shall ensure that –

(a) immediate steps are taken to –

(i) mitigate the effects of the event,

(ii) restore the situation to normal, and

(iii) inform any person who may be affected; and

(b) only those persons who are responsible for the carrying out of repairs and other necessary work are permitted in the affected area and they are provided with –

(i) appropriate respiratory protective equipment and protective clothing, and

(ii) any necessary specialised safety equipment and plant,

which shall be used until the situation is restored to normal.

Duty to prevent or reduce the spread of asbestos

15. Every employer shall prevent or, where this is not reasonably practicable, reduce to the lowest level reasonably practicable, the spread of asbestos from any place where work under his control is carried out.

Cleanliness of premises and plant

16.– (1) Every employer who undertakes work which exposes or is liable to expose his employees to asbestos shall ensure that –

(a) the premises, or those parts of premises where that work is carried out, and the plant used in connection with that work are kept in a clean state; and

(b) where such work has been completed, the premises, or those parts of the premises where the work was carried out, are thoroughly cleaned.

(2) Subject to paragraph (3), where a manufacturing process which gives rise to asbestos dust is carried out in a building, the employer shall ensure that any part of the building in which the process is carried out is –

(a) so designed and constructed as to facilitate cleaning; and

(b) is equipped with an adequate and suitable vacuum cleaning system which shall, where reasonably practicable, be a fixed system.

(3) Paragraph (2)(a) shall not apply to a building in which, prior to 1 March 1988, there was carried out a process to which either –

(a) as then in force, regulation 13 of the Asbestos Regulations 1969[7] applied and the process was carried out in compliance with that regulation; or

(b) that regulation did not apply.

Designated areas

17.– (1) Every employer shall ensure that any area in which work under his control is carried out is designated as –

(a) an asbestos area, where the exposure to asbestos of an employee who worked in that area for the whole of his working time would exceed or would be liable to exceed the action level; and

(b) a respirator zone, where the concentration of asbestos in that area would exceed or would be liable to exceed a relevant control limit.

(2) Asbestos areas and respirator zones shall be clearly and separately demarcated and identified by notices indicating –

(a) that the area is an asbestos area or a respirator zone or both, as the case may be; and

(b) in the case of a respirator zone, that the exposure of an employee who enters it is liable to exceed a relevant control limit and, if it does, that respiratory protective equipment must be worn.

(3) The employer shall not permit any employee, other than an employee who by reason of his work is required to be in an

[7] S.I. 1969/690 – revoked by S.I. 1987/2115.

area designated in accordance with paragraph (1), to enter or remain in any such area and only employees who are so permitted shall enter or remain in any such area.

(4) Every employer shall ensure that –

(a) his employees do not eat, drink or smoke in an area designated as an asbestos area or a respirator zone; and

(b) arrangements are made for such employees to eat or drink in some other place.

Air monitoring

18.– (1) Subject to paragraph (2), every employer shall monitor the exposure of his employees to asbestos by measurement of asbestos fibres present in the air –

(a) at regular intervals; and

(b) when a change occurs which may affect that exposure.

(2) Paragraph (1) shall not apply where –

(a) the exposure of an employee is not liable to exceed the action level; or

(b) the employer is able to demonstrate by another method of evaluation that the requirements of regulation 10(1) have been complied with.

(3) The employer shall keep a suitable record of –

(a) monitoring carried out in accordance with paragraph (1); or

(b) where he decides that monitoring is not required because paragraph (2)(b) applies, the reason for that decision.

(4) The record required by paragraph (3), or a suitable summary thereof, shall be kept –

(a) in a case where exposure is such that a health record is required to be kept under regulation 21, for at least 40 years; or

(b) in any other case, for at least 5 years,

from the date of the last entry made in it.

(5) In relation to the record required by paragraph (3), the employer shall –

(a) on reasonable notice being given, allow an employee access to his personal monitoring record;

(b) provide the Executive with copies of such monitoring records as the Executive may require; and

(c) if he ceases to trade, notify the Executive forthwith in writing and make available to the Executive all monitoring records kept by him.

Standards for air testing

19.– (1) Every employer who carries out any measurement of the concentration of asbestos fibres present in the air shall ensure that he meets criteria equivalent to those set out in the paragraphs of ISO 17025 which cover organisation, quality systems, control of records, personnel, accommodation and environmental conditions, test and calibration methods, method validation, equipment, handling of test and calibration items, and reporting results.

(2) Every employer who requests a person to carry out any measurement of the concentration of asbestos fibres present in the air shall ensure that that person is accredited by an appropriate body as complying with ISO 17025.

Standards for analysis

20.– (1) Every employer who analyses a sample of any material to determine whether it contains asbestos shall ensure that he meets criteria equivalent to those set out in the paragraphs of ISO 17025 which cover organisation, quality systems, control of records, personnel, accommodation and environmental conditions, test and calibration methods, method validation, equipment, handling of test and calibration items, and reporting results.

(2) Every employer who requests a person to analyse a sample of any material to determine whether it contains asbestos shall ensure that that person is accredited by an appropriate body as complying with ISO 17025.

Health records and medical surveillance

21.– (1) Every employer shall ensure that –

(a) a health record, containing particulars approved by the Executive, relating to each of his employees who is exposed to asbestos is maintained unless the exposure of that employee does not exceed the action level; and

(b) that record or a copy thereof is kept available in a suitable form for at least 40 years from the date of the last entry made in it.

(2) Every employer shall ensure that each of his employees who is exposed to asbestos is under adequate medical surveillance by a relevant doctor unless the exposure of that employee does not exceed the action level.

(3) The medical surveillance required by paragraph (2) shall include –

(a) a medical examination not more than 2 years before the beginning of such exposure; and

(b) periodic medical examinations at intervals of not more than 2 years or such shorter time as the relevant doctor may require while such exposure continues,

and each such medical examination shall include a specific examination of the chest.

(4) Where an employee has been examined in accordance with paragraph (3), the relevant doctor shall issue a certificate to the employer and employee stating –

(a) that the employee has been so examined; and

(b) the date of the examination,

and the employer shall keep that certificate or a copy thereof for at least 40 years from the date on which it was issued.

(5) An employee to whom this regulation applies shall, when required by his employer and at the cost of the employer, present himself during his working hours for such examination and tests as may be required for the purposes of paragraph (3) and shall furnish the relevant doctor with such information concerning his health as the relevant doctor may reasonably require.

(6) Where, for the purpose of carrying out his functions under these Regulations, a relevant doctor requires to inspect any record kept for the purposes of these Regulations, the employer shall permit him to do so.

(7) Where medical surveillance is carried out on the premises of the employer, the employer shall ensure that suitable facililities are made available for the purpose.

(8) The employer shall –

(a) on reasonable notice being given, allow an employee access to his personal health record;

(b) provide the Executive with copies of such personal health records as the Executive may require; and

(c) if he ceases to trade, notify the Executive forthwith in writing and make available to the Executive all personal health records kept by him.

(9) Where, as a result of medical surveillance, an employee is found to have an identifiable disease or adverse health effect which is considered by a relevant doctor to be the result of exposure to asbestos at work the employer of that employee shall –

(a) ensure that a suitable person informs the employee accordingly and provides the employee with information and advice regarding further medical surveillance;

(b) review the risk assessment;

(c) review any measure taken to comply with regulation 10, taking into account any advice given by a relevant doctor or by the Executive;

(d) consider assigning the employee to alternative work where there is no risk of further exposure to asbestos, taking into account any advice given by a relevant doctor; and

(e) provide for a review of the health of every other employee who has been similarly exposed, including a medical examination (which shall include a specific examination of the chest) where such an examination is recommended by a relevant doctor or by the Executive.

Washing and changing facilities

22.– (1) Every employer shall ensure that, for any of his employees who is exposed or liable to be exposed to asbestos, there are provided –

(a) adequate washing and changing facilities;

(b) where he is required to provide protective clothing, adequate facilities for the storage of –

(i) that protective clothing, and

(ii) personal clothing not worn during working hours; and

(c) where he is required to provide respiratory protective equipment, adequate facilities for the storage of that equipment.

(2) The facilities provided under paragraph (1) for the storage of –

(a) personal protective clothing;

(b) personal clothing not worn during working hours; and

(c) respiratory protective equipment,

shall be separate from each other.

Storage, distribution and labelling of raw asbestos and asbestos waste

23.– (1) Every employer who undertakes work with asbestos shall ensure that raw asbestos or waste which contains asbestos is not –

(a) stored;

(b) received into or despatched from any place of work; or

(c) distributed within any place of work, except in a totally enclosed distribution system,

unless it is in a sealed container clearly marked in accordance with paragraphs (2) and (3) showing that it contains asbestos.

(2) Raw asbestos shall be labelled in accordance with the provision of Schedule 2.

(3) Waste containing asbestos shall be labelled –

(a) where the Carriage of Dangerous Goods (Classification, Packaging and Labelling) and Use of Transportable Pressure Receptacles Regulations 1996[8] apply, in accordance with those Regulations;

(b) where the waste is conveyed by road in a road tanker or tank container in circumstances where the Carriage of Dangerous Goods by Road Regulations 1996[9] apply, in accordance with those Regulations; and

(c) in any other case, in accordance with the provisions of Schedule 2.

Supply of products containing asbestos for use at work

24.– (1) Subject to paragraph (2), a person shall not supply a product which contains asbestos, being an article or substance for use at work, unless that product is labelled in accordance with the provisions of Schedule 2.

(2) Where a component of a product contains asbestos, it shall be sufficient compliance with this regulation if that

[8] S.I. 1996/2092, amended by S.I. 1998/2885, 1999/303, 2001/1426 and 2002/1689.

[9] S.I. 1996/2095, amended by S.I. 1998/2885 and 1999/303.

component is labelled in accordance with the provisions of Schedule 2 except that where the size of that component makes it impossible for a label to be fixed to it neither that component nor the product need be labelled.

Exemption certificates

25.- (1) Subject to paragraph (2), the Executive may, by a certificate in writing, exempt any person or class of persons from all or any of the requirements or prohibitions imposed by regulations 4, 5, 11, 12, 20 and 21(5) to (7) of these Regulations and any such exemption may be granted subject to conditions and to a limit of time and may be revoked by a further certificate in writing at any time.

(2) The Executive shall not grant any such exemption unless having regard to the circumstances of the case and in particular to –

(a) the conditions, if any, which it proposes to attach to the exemption; and

(b) any other requirements imposed by or under any enactments which apply to the case,

it is satisfied that the health and safety of persons who are likely to be affected by the exemption will not be prejudiced in consequence of it.

Extension outside Great Britain

26. These Regulations shall apply to any work outside Great Britain to which sections 1 to 59 and 80 to 82 of the Health and Safety at Work etc. Act 1974 apply by virtue of the Health and Safety at Work etc. Act 1974 (Application Outside Great Britain) Order 2001[10] as they apply to work in Great Britain.

Revocations, amendments and savings

27.- (1) The following Regulations are revoked –

(a) the Control of Asbestos at Work Regulations 1987[11];

(b) the Control of Asbestos at Work (Amendment) Regulations 1992[12]; and

(c) the Control of Asbestos at Work (Amendment) Regulations 1998[13].

---

[10] S.I. 2001/2127.

[11] S.I. 1987/2115.

[12] S.I. 1992/3068.

[13] S.I. 1998/3235.

(2) After paragraph 4 of Schedule 2 to the Health and Safety (Enforcing Authority) Regulations 1998[14], insert the following paragraph –

"4A. The reference in paragraph 4(a)(iii) to a physically segregated area does not include an area segregated only in order to prevent the escape of asbestos; and in this paragraph "asbestos" has the meaning assigned to it by regulation 2(1) of the Control of Asbestos at Work Regulations 2002[15].".

(3) Any record or register required to be kept under any Regulations revoked by paragraph (1) shall, notwithstanding that revocation, be kept in the same manner and for the same period as specified in those Regulations as if these Regulations had not been made, except that the Executive may approve the keeping of records at a place or in a form other than at the place where, or in the form in which, records were required to be kept under the Regulations so revoked.

Defence

28. Subject to regulation 21 of the Management of Health and Safety at Work Regulations 1999[16], in any proceedings for an offence consisting of a contravention of these Regulations it shall be a defence for any person to prove that he took all reasonable precautions and exercised all due diligence to avoid the commission of that offence.

Signed by authority of the Secretary of State.

N Brown
Minister of State, Department for Work and Pensions

24th October 2002

SCHEDULE 1

Regulation 8(2)

PARTICULARS TO BE INCLUDED IN A NOTIFICATION

The following particulars are to be included in a notification made in accordance with regulation 8(2), namely –

(a) the name of the notifier and the address and telephone number of his usual place of business;

[14] S.I. 1998/494.

[15] S.I. 2002/2675.

[16] S.I. 1999/3242.

(b) a brief description of –

(i) the type(s) of asbestos to be used or handled (crocidolite, amosite, chrysotile or other),

(ii) the maximum quantity of asbestos to be held at any one time on the premises at which the work is to take place,

(iii) the activities or processes involved, and

(iv) the products to be manufactured, where applicable; and

(c) the date of commencement of the work activity.

SCHEDULE 2

Regulations 13(4), 23(2) and (3) and 24

THE LABELLING OF RAW ASBESTOS, ASBESTOS WASTE AND PRODUCTS CONTAINING ASBESTOS

1.– (1) Subject to sub-paragraphs (2) and (3) of this paragraph, the label to be used on –

(a) raw asbestos (together with the labelling required under the Chemicals (Hazard Information and Packaging for Supply) Regulations 2002[17] and the Carriage of Dangerous Goods (Classification, Packaging and Labelling) and Use of Transportable Pressure Receptacles Regulations 1996);

(b) asbestos waste, when required to be so labelled by regulation 23(3); and

(c) products containing asbestos, including used protective clothing to which regulation 13(2) applies,

shall be in the form and in the colours of the following diagram and shall comply with the specifications set out in paragraphs 2 and 3.

[17] S.I. 2002/1689.

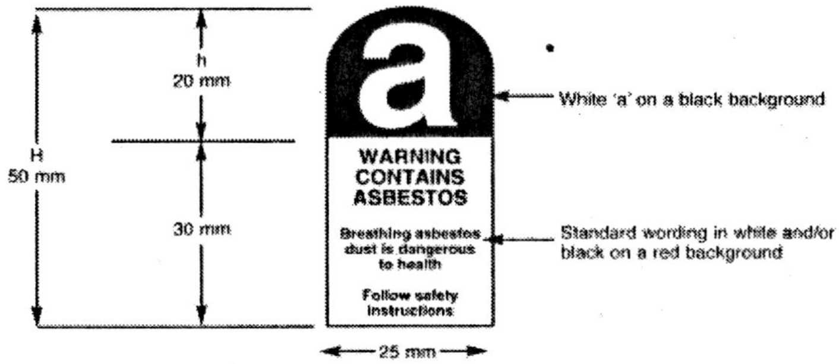

(2) In the case of a product containing crocidolite, the words "contains asbestos" shown in the diagram shall be replaced by the words "contains crocidolite/blue asbestos".

(3) Where the label is printed directly onto a product, a single colour contrasting with the background colour may be used.

2. The dimensions in millimetres of the label referred to in paragraph 1(1) shall be those shown on the diagram in that paragraph, except that larger measurements may be used, but in that case the dimension indicated as h on the diagram shall be 40% of the dimension indicated as H.

3. The label shall be clearly and indelibly printed so that the words in the lower half of the label can be easily read, and those words shall be printed in black or white.

4.– (1) Where a product containing asbestos may undergo processing or finishing it shall bear a label containing safety instructions appropriate to the particular product and in particular the following instructions –

"operate if possible out of doors or in a well ventilated place";

"preferably use hand tools or low speed tools equipped, if necessary, with an appropriate dust extraction facility. If high speed tools are used, they should always be so equipped";

"if possible, dampen before cutting or drilling"; and

"dampen dust, place it in a properly closed receptacle and dispose of it safely".

(2) Additional safety information given on a label shall not detract from or contradict the safety information given in accordance with sub-paragraph (1).

5.– (1) Labelling of packaged and unpackaged products containing asbestos in accordance with the foregoing paragraphs shall be effected by means of –

(a) an adhesive label firmly affixed to the product or its packaging, as the case may be;

(b) a tie-on label firmly attached to the product or its packaging, as the case may be; or

(c) direct printing onto the product or its packaging, as the case may be.

(2) Where, in the case of an unpackaged product containing asbestos, it is not reasonably practicable to comply with the provisions of sub-paragraph (1) the label shall be printed on a suitable sheet accompanying the product.

(3) Labelling of raw asbestos and asbestos waste shall be effected in accordance with sub-paragraph (1)(a) or (c).

(4) For the purposes of this Schedule but subject to sub-paragraph (5), a product supplied in loose plastic or other similar wrapping (including plastic and paper bags) but no other packaging, shall be treated as being supplied in a package whether the product is placed in such wrapping at the time of its supply or was already so wrapped previously.

(5) No wrapping in which a product is placed at the time of its supply shall be regarded as packaging if any product contained in it is labelled in accordance with the requirements of this Schedule or any other packaging in which that product is contained is so labelled.

EXPLANATORY NOTE

(This note is not part of the Regulations)

1. These Regulations re-enact, with modifications, the Control of Asbestos at Work Regulations 1987 (S.I. 1987/2115) ("the 1987 Regulations") as amended. The 1987 Regulations imposed requirements for the protection of employees who might be exposed to asbestos at work and of other persons who might be affected by such work and also imposed certain duties on employees concerning their own protection from such exposure.

2. The Regulations, with the exception of regulations 4, 5, 11, 12, 20 and 21(5) to (7), implement as respects Great Britain –

(a) Council Directive 76/769/EEC ("the Marketing and Use Directive") (OJ No L 262, 27.9.76, p.201) as amended by

Council Directive 83/478/EEC (OJ No L 263, 24.9.83, p.33) on the approximation of the laws, regulations and administrative provisions of the Member States relating to restrictions on the marketing and use of certain dangerous substances and preparations, insofar as it relates to the labelling of products for use at work containing asbestos;

(b) Council Directive 83/477/EEC (OJ No L 263, 24.9.83, p.25) as amended by Council Directive 91/382/EEC (OJ No L 206, 29.7.91, p.16) on the protection of workers from the risks related to exposure to asbestos at work except insofar as this Directive was implemented by the Asbestos (Prohibitions) Regulations 1992 (S.I. 1992/3067) which prohibited asbestos spraying;

(c) Council Directive 90/394/EEC (OJ No L 196, 26.7.90, p.38) on the protection of workers from the risks related to exposure to carcinogens at work (sixth individual Directive within the meaning of Article 16(1) of Directive 89/391/EEC) insofar as it relates to asbestos; and

(d) Council Directive 98/24/EC (OJ No L 131, 5.5.98, p.11) on the protection of the health and safety of workers from risks related to chemical agents at work (fourteenth individual Directive within the meaning of Article 16(1) of Directive 89/391/EEC) insofar as it relates to risks to health from exposure to asbestos.

3. In addition to minor and drafting amendments, the Regulations make the following changes of substance –

(a) include certain further definitions (regulation 2);

(b) introduce a duty to manage asbestos in non-domestic premises. Persons with repair and maintenance responsibilities for the premises will need to ensure that asbestos-containing materials within those premises are properly managed, and that information about the location and condition of the materials is passed on to those likely to disturb them (regulation 4);

(c) extend the matters to be considered when carrying out an assessment of the risk from exposure to asbestos and require the risk assessment to be reviewed when the results of monitoring show it to be necessary (regulation 6);

(d) introduce a duty on the employer to prepare procedures, provide information and establish warning systems to deal with an emergency in the workplace related to the use of asbestos in a work process or the removal or repair of asbestos-containing materials (regulation 14).

(e) provide that air monitoring must occur unless the exposure of an employee to asbestos is not likely to exceed the action level or the employer can demonstrate in another way that the requirements in regulation 10(1) have been complied with (regulation 18);

(f) require any person analysing a material in order to determine whether it contains asbestos to comply with European Standard EN ISO/IEC 17025 (regulation 20);

(g) extend the duties on employers with respect to health records and medical surveillance where an employee is found to have an identifiable disease or adverse health effect caused by exposure to asbestos at work (regulation 21); and

(h) amend the Health and Safety (Enforcing Authority) Regulations 1998 (S.I. 1998/494) in order to clarify that the Executive will not be the enforcing authority solely because an area has been segregated in order to prevent the spread of asbestos (regulation 27).

4. European Standard EN ISO/IEC 17025, "General requirements for the competence of testing and calibration laboratories" accepted by CEN/CENELEC on 6th December 1999 can be obtained from the British Standards Institution, BSI House, 389 Chiswick High Road, London W4 4AL.

5. A copy of the regulatory impact assessment prepared in respect of these Regulations can be obtained from the Health and Safety Executive, Economic Advisers Unit, Rose Court, 2 Southwark Bridge, London SE1 9HS. A copy of the transposition note in relation to implementation of the Directives set out in paragraph 2 can be obtained from the Health and Safety Executive, International Branch at the same address. Copies of both these documents have been placed in the Library of each House of Parliament.

APPENDIX 5
THE CONTROL OF SUBSTANCES HAZARDOUS TO HEALTH REGULATIONS 2002

STATUTORY INSTRUMENTS

2002 No. 2677

HEALTH AND SAFETY

The Control of Substances Hazardous to Health Regulations 2002

Made	*24th October 2002*
Laid before Parliament	*31st October 2002*
Coming into force	*21st November 2002*

ARRANGEMENT OF REGULATIONS

12. Information, instruction and training for persons who may be exposed to substances hazardous to health.
13. Arrangements to deal with accidents, incidents and emergencies.
14. Provisions relating to certain fumigations.
15. Exemption certificates.
16. Exemptions relating to the Ministry of Defence etc.
17. Extension outside Great Britain.
18. Revocation and savings.
19. Extension of meaning of "work".
20. Modification of section 3(2) of the 1974 Act.
21. Defence.

Schedule 1

Other substances and processes to which the definition of "carcinogen" relates.

Schedule 2

Prohibition of certain substances hazardous to health for certain purposes.

Schedule 3

Additional provisions relating to work with biological agents.

Schedule 4

Frequency of thorough examination and test of local exhaust ventilation plant used in certain processes.

Schedule 5

Specific substances and processes for which monitoring is required.

Schedule 6

Medical surveillance.

Schedule 7

Legislation concerned with the labelling of containers and pipes.

Schedule 8

Fumigations excepted from regulation 14.

Schedule 9

Notification of certain fumigations.

The Secretary of State being the Minister designated[1] for the purpose of section 2(2) of the European Communities Act 1972[2] in relation to the abolition of restrictions on the import or export of goods, in the exercise of the powers conferred on him by the said section 2(2) and sections 15(1), (2), (3)(b), (4), (5)(b), (6)(b) and (9), 52(2) and (3) and 82(3)(a) of, and paragraphs 1(1) to (4), 2, 6(1), 8, 9, 11, 14, 15(1), 16 and 20 of Schedule 3 to, the Health and Safety at Work etc. Act 1974[3] ("the 1974 Act") and of all other powers enabling him in that behalf and for the purpose of giving effect without modifications to proposals submitted to him by the Health and Safety Commission under section 11(2)(d) of the 1974 Act after the carrying out by the said Commission of consultations in accordance with section 50(3) of that Act, hereby makes the following Regulations:

Citation and commencement

1. These Regulations may be cited as the Control of Substances Hazardous to Health Regulations 2002 and shall come into force on 21st November 2002.

Interpretation

2.– (1) In these Regulations –

"the 1974 Act" means the Health and Safety at Work etc. Act 1974;

"the Agreement" means the Agreement on the European Economic Area signed at Oporto on 2nd May 1992 as adjusted by the Protocol signed at Brussels on 17th March 1993[4] and adopted as respects Great Britain by the European Economic Area Act 1993[5];

"appointed doctor" means a registered medical practitioner appointed for the time being in writing by the Executive for the purpose of these Regulations;

"approved" means approved for the time being in writing;

---

[1] S.I. 1992/2661.

[2] 1972 c. 68, the definition of the Treaties referred to in section 2(2) of the European Communities Act 1972 was extended by section 1 of the European Economic Area Act 1993 (c. 51).

[3] 1974 c. 37, sections 11(2), 15(1) and 50(3) were amended by the Employment Protection Act 1975 (c. 71), Schedule 15, paragraphs 4, 6 and 16(3) respectively.

[4] The Agreement was amended by Decision 7/94 of the EEA Joint Committee of 21st March 1994 (OJ No. L160, 28.6.94, p. 1). There are other amendments to the Agreement not relevant to these Regulations.

[5] 1993 c. 51.

"approved classification" of a biological agent means the classification of that agent approved by the Health and Safety Commission;

"approved supply list" has the meaning assigned to it in regulation 2(1) of the CHIP Regulations;

"biological agent" means a micro-organism, cell culture, or human endoparasite, whether or not genetically modified, which may cause infection, allergy, toxicity or otherwise create a hazard to human health;

"carcinogen" means –

(a) a substance or preparation which if classified in accordance with the classification provided for by regulation 4 of the CHIP Regulations would be in the category of danger, carcinogenic (category 1) or carcinogenic (category 2) whether or not the substance or preparation would be required to be classified under those Regulations; or

(b) a substance or preparation –

  (i) listed in Schedule 1, or

  (ii) arising from a process specified in Schedule 1 which is a substance hazardous to health;

"cell culture" means the in-vitro growth of cells derived from multicellular organisms;

"the CHIP Regulations" means the Chemicals (Hazard Information and Packaging for Supply) Regulations 2002[6];

"control measure" means a measure taken to reduce exposure to a substance hazardous to health (including the provision of systems of work and supervision, the cleaning of workplaces, premises, plant and equipment, the provision and use of engineering controls and personal protective equipment);

"employment medical adviser" means an employment medical adviser appointed under section 56 of the Health and Safety at Work etc. Act 1974;

"the Executive" means the Health and Safety Executive;

"fumigation" means an operation in which a substance is released into the atmosphere so as to form a gas to control or kill pests or other undesirable organisms and "fumigate" and "fumigant" shall be construed accordingly;

[6] S.I. 2002/1689.

"Group", in relation to a biological agent, means one of the four hazard Groups specified in paragraph 2 of Schedule 3 to which that agent is assigned;

"hazard", in relation to a substance, means the intrinsic property of that substance which has the potential to cause harm to the health of a person, and "hazardous" shall be construed accordingly;

"health surveillance" means assessment of the state of health of an employee, as related to exposure to substances hazardous to health, and includes biological monitoring;

"inhalable dust" means airborne material which is capable of entering the nose and mouth during breathing, as defined by BS EN 481 1993;

"maximum exposure limit" for a substance hazardous to health means the maximum exposure limit approved by the Health and Safety Commission for that substance in relation to the specified reference period when calculated by a method approved by the Health and Safety Commission;

"medical examination" includes any laboratory tests and X-rays that a relevant doctor may require;

"member State" means a State which is a Contracting Party to the Agreement;

"micro-organism" means a microbiological entity, cellular or non-cellular, which is capable of replication or of transferring genetic material;

"mine" has the meaning assigned to it by section 180 of the Mines and Quarries Act 1954[7];

"occupational exposure standard" for a substance hazardous to health means the standard approved by the Health and Safety Commission for that substance in relation to the specified reference period when calculated by a method approved by the Health and Safety Commission;

"personal protective equipment" means all equipment (including clothing) which is intended to be worn or held by a person at work and which protects that person against one or more risks to his health, and any addition or accessory designed to meet that objective;

"preparation" means a mixture or solution of two or more substances;

---

[7] 1954 c. 70, section 180 was modified by S.I. 1974/2013, 1993/1897 and 1999/2024.

"public road" means (in England and Wales) a highway maintainable at the public expense within the meaning of section 329 of the Highways Act 1980[8] and (in Scotland) a public road within the meaning assigned to that term by section 151 of the Roads (Scotland) Act 1984[9];

"registered dentist" has the meaning assigned to it in section 53(1) of the Dentists Act 1984[10];

"relevant doctor" means an appointed doctor or an employment medical adviser;

"respirable dust" means airborne material which is capable of penetrating to the gas exchange region of the lung, as defined by BS EN 481 1993;

"risk", in relation to the exposure of an employee to a substance hazardous to health, means the likelihood that the potential for harm to the health of a person will be attained under the conditions of use and exposure and also the extent of that harm;

"the risk assessment" means the assessment of risk required by regulation 6(1)(a);

"safety data sheet" means a safety data sheet within the meaning of regulation 5 of the CHIP Regulations;

"substance" means a natural or artificial substance whether in solid or liquid form or in the form of a gas or vapour (including micro-organisms);

"substance hazardous to health" means a substance (including a preparation) –

(a) which is listed in Part I of the approved supply list as dangerous for supply within the meaning of the CHIP Regulations and for which an indication of danger specified for the substance is very toxic, toxic, harmful, corrosive or irritant;

(b) for which the Health and Safety Commission has approved a maximum exposure limit or an occupational exposure standard;

(c) which is a biological agent;

(d) which is dust of any kind, except dust which is a substance within paragraph (a) or (b) above, when present at a concentration in air equal to or greater than –

(i) 10 mg/m^3, as a time-weighted average over an 8-hour period, of inhalable dust, or

[8] 1980 c. 66.

[9] 1984 c. 54.

[10] 1984 c. 24.

(ii) 4 mg/m^3, as a time-weighted average over an 8-hour period, of respirable dust;

(e) which, not being a substance falling within sub-paragraphs (a) to (d), because of its chemical or toxicological properties and the way it is used or is present at the workplace creates a risk to health;

"workplace" means any premises or part of premises used for or in connection with work, and includes –

(a) any place within the premises to which an employee has access while at work; and

(b) any room, lobby, corridor, staircase, road or other place –

(i) used as a means of access to or egress from that place of work, or

(ii) where facilities are provided for use in connection with that place of work,

other than a public road.

(2) In these Regulations, a reference to an employee being exposed to a substance hazardous to health is a reference to the exposure of that employee to a substance hazardous to health arising out of or in connection with work at the workplace.

(3) Where a biological agent has an approved classification, any reference in these Regulations to a particular Group in relation to that agent shall be taken as a reference to the Group to which that agent has been assigned in that approved classification.

Duties under these Regulations

3.– (1) Where a duty is placed by these Regulations on an employer in respect of his employees, he shall, so far as is reasonably practicable, be under a like duty in respect of any other person, whether at work or not, who may be affected by the work carried out by the employer except that the duties of the employer –

(a) under regulation 11 (health surveillance) shall not extend to persons who are not his employees; and

(b) under regulations 10, 12(1) and (2) and 13 (which relate respectively to monitoring, information and training and dealing with accidents) shall not extend to persons who are not his employees, unless those persons are on the premises where the work is being carried out.

(2) These Regulations shall apply to a self-employed person as they apply to an employer and an employee and as if that self-employed person were both an employer and an employee, except that regulations 10 and 11 shall not apply to a self-employed person.

(3) The duties imposed by these Regulations shall not extend to the master or crew of a sea-going ship or to the employer of such persons in relation to the normal shipboard activities of a ship's crew under the direction of the master.

Prohibitions relating to certain substances

4.– (1) Those substances described in Column 1 of Schedule 2 are prohibited to the extent set out in the corresponding entry in Column 2 of that Schedule.

(2) The importation into the United Kingdom, other than from another member State, of the following substances and articles is prohibited, namely –

(a) 2-naphthylamine, benzidine, 4-aminodiphenyl, 4-nitrodiphenyl, their salts and any substance containing any of those compounds in a total concentration equal to or greater than 0.1 per cent by mass;

(b) matches made with white phosphorus,

and a contravention of this paragraph shall be punishable under the Customs and Excise Management Act 1979[11] and not as a contravention of a health and safety regulation.

(3) A person shall not supply during the course of or for use at work a substance or article specified in paragraph (2).

(4) A person shall not supply during the course of or for use at work, benzene or a substance containing benzene unless its intended use is not prohibited by item 11 of Schedule 2.

Application of regulations 6 to 13

5.– (1) Regulations 6 to 13 shall have effect with a view to protecting persons against a risk to their health, whether immediate or delayed, arising from exposure to substances hazardous to health except –

(a) where and to the extent that the following Regulations apply, namely –

(i) the Coal Mines (Respirable Dust) Regulations 1975[12],

(ii) the Control of Lead at Work Regulations 2002[13],

[11] 1979 c. 2.

[12] S.I. 1975/1433, as amended by S.I. 1978/807.

[13] S.I. 2002/2676.

(iii) the Control of Asbestos at Work Regulations 2002[14];

(b) where the substance is hazardous to health solely by virtue of its radioactive, explosive or flammable properties, or solely because it is at a high or low temperature or a high pressure;

(c) where the risk to health is a risk to the health of a person to whom the substance is administered in the course of his medical treatment.

(2) In paragraph (1)(c) "medical treatment" means medical or dental examination or treatment which is conducted by, or under the direction of a –

(a) registered medical practitioner;

(b) registered dentist; or

(c) other person who is an appropriate practitioner for the purposes of section 58 of the Medicines Act 1968[15],

and includes any such examination or treatment conducted for the purpose of research.

Assessment of the risk to health created by work involving substances hazardous to health

6.– (1) An employer shall not carry out work which is liable to expose any employees to any substance hazardous to health unless he has –

(a) made a suitable and sufficient assessment of the risk created by that work to the health of those employees and of the steps that need to be taken to meet the requirements of these Regulations; and

(b) implemented the steps referred to in sub-paragraph (a).

(2) The risk assessment shall include consideration of –

(a) the hazardous properties of the substance;

(b) information on health effects provided by the supplier, including information contained in any relevant safety data sheet;

(c) the level, type and duration of exposure;

(d) the circumstances of the work, including the amount of the substance involved;

[14] S.I. 2002/2675.

[15] 1968 c. 67.

(e) activities, such as maintenance, where there is the potential for a high level of exposure;

(f) any relevant occupational exposure standard, maximum exposure limit or similar occupational exposure limit;

(g) the effect of preventive and control measures which have been or will be taken in accordance with regulation 7;

(h) the results of relevant health surveillance;

(i) the results of monitoring of exposure in accordance with regulation 10;

(j) in circumstances where the work will involve exposure to more than one substance hazardous to health, the risk presented by exposure to such substances in combination;

(k) the approved classification of any biological agent; and

(l) such additional information as the employer may need in order to complete the risk assessment.

(3) The risk assessment shall be reviewed regularly and forthwith if –

(a) there is reason to suspect that the risk assessment is no longer valid;

(b) there has been a significant change in the work to which the risk assessment relates; or

(c) the results of any monitoring carried out in accordance with regulation 10 show it to be necessary,

and where, as a result of the review, changes to the risk assessment are required, those changes shall be made.

(4) Where the employer employs 5 or more employees, he shall record –

(a) the significant findings of the risk assessment as soon as is practicable after the risk assessment is made; and

(b) the steps which he has taken to meet the requirements of regulation 7.

Prevention or control of exposure to substances hazardous to health

7.– (1) Every employer shall ensure that the exposure of his employees to substances hazardous to health is either prevented or, where this is not reasonably practicable, adequately controlled.

(2) In complying with his duty of prevention under paragraph (1), substitution shall by preference be undertaken, whereby the employer shall avoid, so far as is reasonably practicable, the use of a substance hazardous to health at the workplace by replacing it with a substance or process which, under the conditions of its use, either eliminates or reduces the risk to the health of his employees.

(3) Where it is not reasonably practicable to prevent exposure to a substance hazardous to health, the employer shall comply with his duty of control under paragraph (1) by applying protection measures appropriate to the activity and consistent with the risk assessment, including, in order of priority –

(a) the design and use of appropriate work processes, systems and engineering controls and the provision and use of suitable work equipment and materials;

(b) the control of exposure at source, including adequate ventilation systems and appropriate organisational measures; and

(c) where adequate control of exposure cannot be achieved by other means, the provision of suitable personal protective equipment in addition to the measures required by sub-paragraphs (a) and (b).

(4) The measures referred to in paragraph (3) shall include –

(a) arrangements for the safe handling, storage and transport of substances hazardous to health, and of waste containing such substances, at the workplace;

(b) the adoption of suitable maintenance procedures;

(c) reducing, to the minimum required for the work concerned –

(i) the number of employees subject to exposure,

(ii) the level and duration of exposure, and

(iii) the quantity of substances hazardous to health present at the workplace;

(d) the control of the working environment, including appropriate general ventilation; and

(e) appropriate hygiene measures including adequate washing facilities.

(5) Without prejudice to the generality of paragraph (1), where it is not reasonably practicable to prevent exposure to a carcinogen, the employer shall apply the following measures in addition to those required by paragraph (3) –

(a) totally enclosing the process and handling systems, unless this is not reasonably practicable;

(b) the prohibition of eating, drinking and smoking in areas that may be contaminated by carcinogens;

(c) cleaning floors, walls and other surfaces at regular intervals and whenever necessary;

(d) designating those areas and installations which may be contaminated by carcinogens and using suitable and sufficient warning signs; and

(e) storing, handling and disposing of carcinogens safely, including using closed and clearly labelled containers.

(6) Without prejudice to the generality of paragraph (1), where it is not reasonably practicable to prevent exposure to a biological agent, the employer shall apply the following measures in addition to those required by paragraph (3) –

(a) displaying suitable and sufficient warning signs, including the biohazard sign shown in Part IV of Schedule 3;

(b) specifying appropriate decontamination and disinfection procedures;

(c) instituting means for the safe collection, storage and disposal of contaminated waste, including the use of secure and identifiable containers, after suitable treatment where appropriate;

(d) testing, where it is necessary and technically possible, for the presence, outside the primary physical confinement, of biological agents used at work;

(e) specifying procedures for working with, and transporting at the workplace, a biological agent or material that may contain such an agent;

(f) where appropriate, making available effective vaccines for those employees who are not already immune to the biological agent to which they are exposed or are liable to be exposed;

(g) instituting hygiene measures compatible with the aim of preventing or reducing the accidental transfer or release of a biological agent from the workplace, including –

(i) the provision of appropriate and adequate washing and toilet facilities, and

(ii) where appropriate, the prohibition of eating, drinking, smoking and the application of cosmetics in working areas where there is a risk of contamination by biological agents; and

(h) where there are human patients or animals which are, or are suspected of being, infected with a Group 3 or 4 biological agent, the employer shall select the most suitable control and containment measures from those listed in Part II of Schedule 3 with a view to controlling adequately the risk of infection.

(7) Without prejudice to the generality of paragraph (1), where there is exposure to a substance for which a maximum exposure limit has been approved, control of exposure shall, so far as the inhalation of that substance is concerned, only be treated as being adequate if the level of exposure is reduced so far as is reasonably practicable and in any case below the maximum exposure limit.

(8) Without prejudice to the generality of paragraph (1), where there is exposure to a substance for which an occupational exposure standard has been approved, control of exposure shall, so far as the inhalation of that substance is concerned, only be treated as being adequate if –

(a) that occupational exposure standard is not exceeded; or

(b) where that occupational exposure standard is exceeded, the employer identifies the reasons for the standard being exceeded and takes appropriate action to remedy the situation as soon as is reasonably practicable.

(9) Personal protective equipment provided by an employer in accordance with this regulation shall be suitable for the purpose and shall –

(a) comply with any provision in the Personal Protective Equipment Regulations 2002[16] which is applicable to that item of personal protective equipment; or

(b) in the case of respiratory protective equipment, where no provision referred to in sub-paragraph (a) applies, be of a type approved or shall conform to a standard approved, in either case, by the Executive.

(10) Without prejudice to the provisions of this regulation, Schedule 3 shall have effect in relation to work with biological agents.

(11) In this regulation, "adequate" means adequate having regard only to the nature of the substance and the nature and degree of exposure to substances hazardous to health and "adequately" shall be construed accordingly.

Use of control measures etc.

8.– (1) Every employer who provides any control measure, other thing or facility in accordance with these Regulations shall take all reasonable steps to ensure that it is properly used or applied as the case may be.

(2) Every employee shall make full and proper use of any control measure, other thing or facility provided in accordance with these Regulations and, where relevant, shall –

(a) take all reasonable steps to ensure it is returned after use to any accommodation provided for it; and

(b) if he discovers a defect therein, report it forthwith to his employer.

Maintenance, examination and testing of control measures

9.– (1) Every employer who provides any control measure to meet the requirements of regulation 7 shall ensure that, where relevant, it is maintained in an efficient state, in efficient working order, in good repair and in a clean condition.

(2) Where engineering controls are provided to meet the requirements of regulation 7, the employer shall ensure that thorough examination and testing of those controls is carried out –

(a) in the case of local exhaust ventilation plant, at least once every 14 months, or for local exhaust ventilation

[16] S.I. 2002/1144.

plant used in conjunction with a process specified in Column 1 of Schedule 4, at not more than the interval specified in the corresponding entry in Column 2 of that Schedule; or

(b) in any other case, at suitable intervals.

(3) Where respiratory protective equipment (other than disposable respiratory protective equipment) is provided to meet the requirements of regulation 7, the employer shall ensure that thorough examination and, where appropriate, testing of that equipment is carried out at suitable intervals.

(4) Every employer shall keep a suitable record of the examinations and tests carried out in accordance with paragraphs (2) and (3) and of repairs carried out as a result of those examinations and tests, and that record or a suitable summary thereof shall be kept available for at least 5 years from the date on which it was made.

(5) Every employer shall ensure that personal protective equipment, including protective clothing, is:

(a) properly stored in a well-defined place;

(b) checked at suitable intervals; and

(c) when discovered to be defective, repaired or replaced before further use.

(6) Personal protective equipment which may be contaminated by a substance hazardous to health shall be removed on leaving the working area and kept apart from uncontaminated clothing and equipment.

(7) The employer shall ensure that the equipment referred to in paragraph (6) is subsequently decontaminated and cleaned or, if necessary, destroyed.

Monitoring exposure at the workplace

10.- (1) Where the risk assessment indicates that –

(a) it is requisite for ensuring the maintenance of adequate control of the exposure of employees to substances hazardous to health; or

(b) it is otherwise requisite for protecting the health of employees,

the employer shall ensure that the exposure of employees to substances hazardous to health is monitored in accordance with a suitable procedure.

(2) Paragraph (1) shall not apply where the employer is able to demonstrate by another method of evaluation that the requirements of regulation 7(1) have been complied with.

(3) The monitoring referred to in paragraph (1) shall take place –

(a) at regular intervals; and

(b) when any change occurs which may affect that exposure.

(4) Where a substance or process is specified in Column 1 of Schedule 5, monitoring shall be carried out at least at the frequency specified in the corresponding entry in Column 2 of that Schedule.

(5) The employer shall ensure that a suitable record of monitoring carried out for the purpose of this regulation is made and maintained and that that record or a suitable summary thereof is kept available –

(a) where the record is representative of the personal exposures of identifiable employees, for at least 40 years; or

(b) in any other case, for at least 5 years,

from the date of the last entry made in it.

(6) Where an employee is required by regulation 11 to be under health surveillance, an individual record of any monitoring carried out in accordance with this regulation shall be made, maintained and kept in respect of that employee.

(7) The employer shall –

(a) on reasonable notice being given, allow an employee access to his personal monitoring record;

(b) provide the Executive with copies of such monitoring records as the Executive may require; and

(c) if he ceases to trade, notify the Executive forthwith in writing and make available to the Executive all monitoring records kept by him.

Health surveillance

11.– (1) Where it is appropriate for the protection of the health of his employees who are, or are liable to be, exposed to a substance hazardous to health, the employer shall ensure that such employees are under suitable health surveillance.

(2) Health surveillance shall be treated as being appropriate where –

(a) the employee is exposed to one of the substances specified in Column 1 of Schedule 6 and is engaged in a process specified in Column 2 of that Schedule, and there is a reasonable likelihood that an identifiable disease or adverse health effect will result from that exposure; or

(b) the exposure of the employee to a substance hazardous to health is such that –

(i) an identifiable disease or adverse health effect may be related to the exposure,

(ii) there is a reasonable likelihood that the disease or effect may occur under the particular conditions of his work, and

(iii) there are valid techniques for detecting indications of the disease or effect,

and the technique of investigation is of low risk to the employee.

(3) The employer shall ensure that a health record, containing particulars approved by the Executive, in respect of each of his employees to whom paragraph (1) applies, is made and maintained and that that record or a copy thereof is kept available in a suitable form for at least 40 years from the date of the last entry made in it.

(4) The employer shall –

(a) on reasonable notice being given, allow an employee access to his personal health record;

(b) provide the Executive with copies of such health records as the Executive may require; and

(c) if he ceases to trade, notify the Executive forthwith in writing and make available to the Executive all health records kept by him.

(5) If an employee is exposed to a substance specified in Schedule 6 and is engaged in a process specified therein, the health surveillance required under paragraph (1) shall include medical surveillance under the supervision of a relevant doctor at intervals of not more than 12 months or at such shorter intervals as the relevant doctor may require.

(6) Where an employee is subject to medical surveillance in accordance with paragraph (5) and a relevant doctor has certified by an entry in the health record of that employee that in his professional opinion that employee should not be engaged in work which exposes him to that substance or that he should only be so engaged under conditions specified in the record, the employer shall not permit the employee to be engaged in such work except in accordance with the conditions, if any, specified in the health record, unless that entry has been cancelled by a relevant doctor.

(7) Where an employee is subject to medical surveillance in accordance with paragraph (5) and a relevant doctor has certified by an entry in his health record that medical surveillance should be continued after his exposure to that substance has ceased, the employer shall ensure that the medical surveillance of that employee is continued in accordance with that entry while he is employed by the employer, unless that entry has been cancelled by a relevant doctor.

(8) An employee to whom this regulation applies shall, when required by his employer and at the cost of the employer, present himself during his working hours for such health surveillance procedures as may be required for the purposes of paragraph (1) and, in the case of an employee who is subject to medical surveillance in accordance with paragraph (5), shall furnish the relevant doctor with such information concerning his health as the relevant doctor may reasonably require.

(9) Where, as a result of health surveillance, an employee is found to have an identifiable disease or adverse health effect which is considered by a relevant doctor or other occupational health professional to be the result of exposure to a substance hazardous to health the employer of that employee shall –

(a) ensure that a suitably qualified person informs the employee accordingly and provides the employee with information and advice regarding further health surveillance;

(b) review the risk assessment;

(c) review any measure taken to comply with regulation 7, taking into account any advice given by a relevant doctor, occupational health professional or by the Executive;

(d) consider assigning the employee to alternative work where there is no risk of further exposure to that substance, taking into account any advice given by a relevant doctor or occupational health professional; and

(e) provide for a review of the health of any other employee who has been similarly exposed, including a medical examination where such an examination is recommended by a relevant doctor, occupational health professional or by the Executive.

(10) Where, for the purpose of carrying out his functions under these Regulations, a relevant doctor requires to inspect any workplace or any record kept for the purposes of these Regulations, the employer shall permit him to do so.

(11) Where an employee or an employer is aggrieved by a decision recorded in the health record by a relevant doctor to suspend an employee from work which exposes him to a substance hazardous to health (or to impose conditions on such work), he may, by an application in writing to the Executive within 28 days of the date on which he was notified of the decision, apply for that decision to be reviewed in accordance with a procedure approved for the purposes of this paragraph by the Health and Safety Commission, and the result of that review shall be notified to the employee and employer and entered in the health record in accordance with the approved procedure.

Information, instruction and training for persons who may be exposed to substances hazardous to health

12.– (1) Every employer who undertakes work which is liable to expose an employee to a substance hazardous to health shall provide that employee with suitable and sufficient information, instruction and training.

(2) Without prejudice to the generality of paragraph (1), the information, instruction and training provided under that paragraph shall include –

(a) details of the substances hazardous to health to which the employee is liable to be exposed including –

(i) the names of those substances and the risk which they present to health,

(ii) any relevant occupational exposure standard, maximum exposure limit or similar occupational exposure limit,

(iii) access to any relevant safety data sheet, and

(iv) other legislative provisions which concern the hazardous properties of those substances;

(b) the significant findings of the risk assessment;

(c) the appropriate precautions and actions to be taken by the employee in order to safeguard himself and other employees at the workplace;

(d) the results of any monitoring of exposure in accordance with regulation 10 and, in particular, in the case of a substance hazardous to health for which a maximum exposure limit has been approved, the employee or his representatives shall be informed forthwith, if the results of such monitoring show that the maximum exposure limit has been exceeded;

(e) the collective results of any health surveillance undertaken in accordance with regulation 11 in a form calculated to prevent those results from being identified as relating to a particular person; and

(f) where employees are working with a Group 4 biological agent or material that may contain such an agent, the provision of written instructions and, if appropriate, the display of notices which outline the procedures for handling such an agent or material.

(3) The information, instruction and training required by paragraph (1) shall be –

(a) adapted to take account of significant changes in the type of work carried out or methods of work used by the employer; and

(b) provided in a manner appropriate to the level, type and duration of exposure identified by the risk assessment.

(4) Every employer shall ensure that any person (whether or not his employee) who carries out work in connection with the employer's duties under these Regulations has suitable and sufficient information, instruction and training.

(5) Where containers and pipes for substances hazardous to health used at work are not marked in accordance with any relevant legislation listed in Schedule 7, the employer shall, without prejudice to any derogations provided for in that legislation, ensure that the contents of those containers and pipes, together with the nature of those contents and any associated hazards, are clearly identifiable.

Arrangements to deal with accidents, incidents and emergencies

13.– (1) Subject to paragraph (4) and without prejudice to the relevant provisions of the Management of Health and Safety at Work Regulations 1999[17], in order to protect the health of his employees from an accident, incident or emergency related to the presence of a substance hazardous to health at the workplace, the employer shall ensure that –

(a) procedures, including the provision of appropriate first-aid facilities and relevant safety drills (which shall be tested at regular intervals), have been prepared which can be put into effect when such an event occurs;

(b) information on emergency arrangements, including –

(i) details of relevant work hazards and hazard identification arrangements, and

(ii) specific hazards likely to arise at the time of an accident, incident or emergency, is available; and

(c) suitable warning and other communication systems are established to enable an appropriate response, including remedial actions and rescue operations, to be made immediately when such an event occurs.

(2) The employer shall ensure that information on the procedures and systems required by paragraph (1)(a) and (c) and the information required by paragraph(1)(b) is –

(a) made available to relevant accident and emergency services to enable those services, whether internal or external to the workplace, to prepare their own response procedures and precautionary measures; and

(b) displayed at the workplace, if this is appropriate.

(3) Subject to paragraph (4), in the event of an accident, incident or emergency related to the presence of a substance hazardous to health at the workplace, the employer shall ensure that –

(a) immediate steps are taken to –

(i) mitigate the effects of the event,

(ii) restore the situation to normal, and

(iii) inform those of his employees who may be affected;

[17] S.I. 1999/3242.

(b) only those persons who are essential for the carrying out of repairs and other necessary work are permitted in the affected area and they are provided with –

(i) appropriate personal protective equipment, and

(ii) any necessary specialised safety equipment and plant,

which shall be used until the situation is restored to normal; and

(c) in the case of an incident or accident which has or may have resulted in the release of a biological agent which could cause severe human disease, as soon as practicable thereafter his employees or their representatives are informed of –

(i) the causes of that incident or accident, and

(ii) the measures taken or to be taken to rectify the situation.

(4) Paragraph (1) and, provided the substance hazardous to health is not a carcinogen or biological agent, paragraph (3) shall not apply where –

(a) the results of the risk assessment show that, because of the quantity of each substance hazardous to health present at the workplace, there is only a slight risk to the health of employees; and

(b) the measures taken by the employer to comply with the duty under regulation 7(1) are sufficient to control that risk.

(5) An employee shall report forthwith, to his employer or to any other employee of that employer with specific responsibility for the health and safety of his fellow employees, any accident or incident which has or may have resulted in the release of a biological agent which could cause severe human disease.

Provisions relating to certain fumigations

14.– (1) This regulation shall apply to fumigations in which the fumigant used or intended to be used is hydrogen cyanide, phosphine or methyl bromide, except that paragraph (2) shall not apply to fumigations using the fumigant specified in Column 1 of Schedule 8 when the nature of the fumigation is that specified in the corresponding entry in Column 2 of that Schedule.

(2) An employer shall not undertake fumigation to which this regulation applies unless he has –

(a) notified the persons specified in Part I of Schedule 9 of his intention to undertake the fumigation; and

(b) provided to those persons the information specified in Part II of that Schedule,

at least 24 hours in advance, or such shorter time in advance as the persons required to be notified may agree.

(3) An employer who undertakes a fumigation to which this regulation applies shall ensure that, before the fumigant is released, suitable warning notices have been affixed at all points of reasonable access to the premises or to those parts of the premises in which the fumigation is to be carried out and that after the fumigation has been completed, and the premises are safe to enter, those warning notices are removed.

Exemption certificates

15.– (1) Subject to paragraph (2) the Executive may, by a certificate in writing, exempt any person or class of persons or any substance or class of substances from all or any of the requirements or prohibitions imposed by regulations 4 (to the extent permitted by article 9 of Council Directive 98/24/EC), 8, 9, 11(8), (10) and (11) and 14 of these Regulations and any such exemption may be granted subject to conditions and to a limit of time and may be revoked by a certificate in writing at any time.

(2) The Executive shall not grant any such exemption unless having regard to the circumstances of the case and, in particular, to –

(a) the conditions, if any, which it proposes to attach to the exemption; and

(b) any requirements imposed by or under any enactments which apply to the case,

it is satisfied that the health and safety of persons who are likely to be affected by the exemption will not be prejudiced in consequence of it.

Exemptions relating to the Ministry of Defence etc.

16.– (1) In this regulation –

(a) "Her Majesty's Forces" means any of the naval, military or air forces of the Crown, whether raised inside or outside the United Kingdom and whether any such force is a regular, auxiliary or reserve force, and includes any civilian employed by those forces;

(b) "visiting force" has the same meaning as it does for the purposes of any provision of Part I of the Visiting Forces Act 1952[18]; and

(c) "headquarters" means a headquarters for the time being specified in Schedule 2 to the Visiting Forces and International Headquarters (Application of Law) Order 1999[19].

(2) The Secretary of State for Defence may, in the interests of national security, by a certificate in writing exempt –

(a) any of Her Majesty's Forces;

(b) any visiting force;

(c) members of a visiting force working in or attached to a headquarters; or

(d) any person engaged in work involving substances hazardous to health, if that person is under the direct supervision of a representative of the Secretary of State for Defence,

from all or any of the requirements or prohibitions imposed by these Regulations and any such exemption may be granted subject to conditions and to a limit of time and may be revoked at any time by a certificate in writing, except that, where any such exemption is granted, suitable arrangements shall be made for the assessment of the health risk created by the work involving substances hazardous to health and for adequately controlling the exposure to those substances of persons to whom the exemption relates.

(3) Regulation 11(11) shall not apply in relation to –

(a) any visiting force; or

(b) members of a visiting force working in or attached to a headquarters.

Extension outside Great Britain

17.– (1) Subject to paragraph (2), these Regulations shall apply to and in relation to any activity outside Great Britain to which sections 1 to 59 and 80 to 82 of the 1974 Act apply by virtue of the Health and Safety at Work etc. Act 1974 (Application outside Great Britain) Order 2001[20] as those provisions apply within Great Britain.

[18] 1952 c. 67.

[19] S.I. 1999/1736.

[20] S.I. 2001/2127.

(2) These Regulations shall not extend to Northern Ireland except insofar as they relate to imports of substances and articles referred to in regulation 4(2) into the United Kingdom.

Revocation and savings

18.– (1) The Control of Substances Hazardous to Health Regulations 1999[21] are revoked.

(2) Any record or register required to be kept under the Regulations revoked by paragraph (1) shall, notwithstanding that revocation, be kept in the same manner and for the same period as specified in those Regulations as if these Regulations had not been made, except that the Executive may approve the keeping of records at a place or in a form other than at the place where, or in the form in which, records were required to be kept under the Regulations so revoked.

Extension of meaning of "work"

19. For the purposes of Part I of the 1974 Act the meaning of "work" shall be extended to include any activity involving the consignment, storage or use of a Group 2, 3 or 4 biological agent and the meaning of "at work" shall be extended accordingly, and in that connection the references to employer in paragraphs 5 and 6 of Schedule 3 include references to any persons carrying out such an activity.

Modification of section 3(2) of the 1974 Act

20. Section 3(2) of the 1974 Act shall be modified in relation to an activity involving the consignment, storage or use of any of the biological agents referred to in regulation 19 so as to have effect as if the reference therein to a self-employed person is a reference to any person who is not an employer or an employee and the reference therein to his undertaking includes a reference to such an activity.

Defence

21. Subject to regulation 21 of the Management of Health and Safety at Work Regulations 1999[22], in any proceedings for an offence consisting of a contravention of these Regulations it shall be a defence for any person to prove that he took all reasonable precautions and exercised all due diligence to avoid the commission of that offence.

Signed by authority of the Secretary of State.

N. Brown
Minister of State, Department for Work and Pensions

24th October 2002

[21] S.I. 1999/437.

[22] S.I. 1999/3242.

SCHEDULE 1

Regulation 2(1)

OTHER SUBSTANCES AND PROCESSES TO WHICH THE DEFINITION OF "CARCINOGEN" RELATES

Aflatoxins.

Arsenic.

Auramine manufacture.

Calcining, sintering or smelting of nickel copper matte or acid leaching or electrorefining of roasted matte.

Coal soots, coal tar, pitch and coal tar fumes.

Hardwood dusts.

Isopropyl alcohol manufacture (strong acid process).

Leather dust in boot and shoe manufacture, arising during preparation and finishing.

Magenta manufacture.

Mustard gas ([beta], [beta]' -dichlorodiethyl sulphide).

Rubber manufacturing and processing giving rise to rubber process dust and rubber fume.

Used engine oils.

SCHEDULE 2

Regulation 4(1)

PROHIBITION OF CERTAIN SUBSTANCES HAZARDOUS TO HEALTH FOR CERTAIN PURPOSES

Column 1
Column 2

Description of substance
Purpose for which the substance is prohibited

1. 2-naphthylamine; benzidine;

   4-aminodiphenyl;

   4-nitrodiphenyl; their salts and any substance containing any of those compounds, in a total concentration equal to or greater than 0.1 per cent by mass.

Manufacture and use for all purposes including any manufacturing process in which a substance described in Column 1 of this item is formed.

2. Sand or other substance containing free silica.

Use as an abrasive for blasting articles in any blasting apparatus.

3. A substance –

(a) containing compounds of silicon calculated as silica to the extent of more than 3 per cent by weight of dry material, other than natural sand, zirconium silicate (zircon), calcined china clay, calcined aluminous fireclay, sillimanite, calcined or fused alumina, olivine; or

(b) composed of or containing dust or other matter deposited from a fettling or blasting process.

Use as a parting material in connection with the making of metal castings.

4. Carbon disulphide.

Use in the cold-cure process of vulcanising in the proofing of cloth with rubber.

5. Oils other than white oil, or oil of entirely animal or vegetable origin or entirely of mixed animal and vegetable origin.

Use for oiling the spindles of self-acting mules.

6. Ground or powdered flint or quartz other than natural sand.

Use in relation to the manufacture or decoration of pottery for the following purposes:

(a) the placing of ware for the biscuit fire;

(b) the polishing of ware;

(c) as the ingredient of a wash for saggars, trucks, bats, cranks, or other articles used in supporting ware during firing; and

(d) as dusting or supporting powder in potters' shops.

7. Ground or powdered flint or quartz other than –

(a) natural sand; or

(b) ground or powdered flint or quartz

which forms parts of a slop or paste.

Use in relation to the manufacture or decoration of pottery for any purpose except –

(a) use in a separate room or building for –

(i) the manufacture of powdered flint or quartz, or

(ii) the making of frits or glazes or the making of colours or coloured slips for the decoration of pottery;

(b) use for the incorporation of the substance into the body of ware in an enclosure in which no person is employed and which is constructed and ventilated to prevent the escape of dust.

8. Dust or powder of a refractory material containing not less than 80 per cent of silica other than natural sand.

Use for sprinkling the moulds of silica bricks, namely bricks or other articles composed of refractory material and containing not less than 80 per cent of silica.

9. White phosphorus.

Use in the manufacture of matches.

10. Hydrogen cyanide.

Use in fumigation except when –

(a) released from an inert material in which hydrogen cyanide is absorbed;

(b) generated from a gassing powder; or

(c) applied from a cylinder through suitable piping and applicators other than for fumigation in the open air to control or kill mammal pests.

11. Benzene and any substance containing benzene in a concentration equal to or greater than 0.1 per cent by mass, other than –

(a) motor fuels covered by Council Directive 85/210/EEC (OJ No. L96, 3.4.85, p. 25);

(b) waste covered by Council Directives 75/442/EEC (OJ No. L194, 25.7.75, p. 39), as amended by Council Directive 91/156/EEC (OJ No. L78, 26.3.91, p. 32), and 91/689/EEC (OJ No. L377, 31.12.91, p. 20).

Use for all purposes except –

(a) use in industrial processes; and

(b) for the purposes of research and development or for the purpose of analysis.

12. The following substances –

Chloroform CAS No. 67–66–3;

Carbon Tetrachloride CAS No. 56–23–5;

1,1,2 Trichloroethane CAS No. 79–00–5;

1,1,2,2 Tetrachloroethane CAS No. 79–34–5;

1,1,1,2 Tetrachloroethane CAS No. 630–20–6;

Pentachloroethane CAS No. 76–01–7;

Vinylidene chloride

(1,1 Dichloroethylene) CAS No. 75–35–4;

1,1,1 Trichloroethane CAS No. 71–55–6,

and any substance containing one or more of those substances in a concentration equal to or greater than 0.1 per cent by mass, other than –

(a) medicinal products;

(b) cosmetic products.

Supply for use at work in diffusive applications such as in surface cleaning and the cleaning of fabrics except for the purposes of research and development or for the purpose of analysis.

In this Schedule –

"aerosol dispenser" means an article which consists of a non-reusable receptacle containing a gas compressed, liquefied or dissolved under pressure, with or without liquid, paste or powder and fitted with a release device allowing the contents to be ejected as solid or liquid particles in suspension in a gas, as a foam, paste or powder or in a liquid state;

"blasting apparatus" means apparatus for cleaning, smoothing, roughening or removing of part of the surface of any article by the use as an abrasive of a jet of sand, metal shot or grit or other material propelled by a blast of compressed air or steam or by a wheel;

"CAS No. " is the number assigned to a substance by the Chemical Abstract Service;

"cosmetic product" has the meaning assigned to it in regulation 2(1) of the Cosmetic Products (Safety) Regulations 1996[23] (including any aerosol dispenser containing a cosmetic product);

"gassing powder" means a chemical compound in powder form which reacts with atmospheric moisture to generate hydrogen cyanide;

[23] S.I. 1996/2925.

"medicinal product" means a substance or preparation which is –

(a) intended for use as a medicinal product within the meaning of section 130 of the Medicines Act 1968[24]; or

(b) a substance or preparation specified in an order made under section 104 or 105 of the Medicines Act 1968 which is for the time being in force and which directs that specified provisions of that Act shall have effect in relation to that substance or preparation as such provisions have effect in relation to medicinal products within the meaning of that Act;

"use as a parting material" means the application of the material to the surface or parts of the surface of a pattern or of a mould so as to facilitate the separation of the pattern from the mould or the separation of parts of the mould;

"white oil" means a refined mineral oil conforming to a specification approved by the Executive and certified by its manufacturer as so conforming.

SCHEDULE 3

Regulation 7(10)

ADDITIONAL PROVISIONS RELATING TO WORK WITH BIOLOGICAL AGENTS

PART I

PROVISIONS OF GENERAL APPLICATION TO BIOLOGICAL AGENTS

Interpretation

1. In this Schedule "diagnostic service" means any activity undertaken solely with the intention of analysing specimens or samples from a human patient or animal in which a biological agent is or is suspected of being present for purposes relating to the assessment of the clinical progress, or assistance in the clinical management, of that patient or animal, and "diagnosis" shall be construed accordingly.

Classification of biological agents

2.– (1) Where a biological agent does not have an approved classification, the employer shall provisionally classify that agent in accordance with sub-paragraph (2), having regard to the nature of the agent and the properties of which he may reasonably be expected to be aware.

[24] 1968 c. 67.

(2) When provisionally classifying a biological agent the employer shall assign that agent to one of the following Groups according to its level of risk of infection and, if in doubt as to which of two alternative Groups is the most appropriate, he shall assign it to the higher of the two –

(a) Group 1 – unlikely to cause human disease;

(b) Group 2 – can cause human disease and may be a hazard to employees; it is unlikely to spread to the community and there is usually effective prophylaxis or treatment available;

(c) Group 3 – can cause severe human disease and may be a serious hazard to employees; it may spread to the community, but there is usually effective prophylaxis or treatment available;

(d) Group 4 – causes severe human disease and is a serious hazard to employees; it is likely to spread to the community and there is usually no effective prophylaxis or treatment available.

(3) Where an employer is using a biological agent which has an approved classification and the risk of infection for that particular agent is different to that expected, the employer shall reclassify the agent in consultation with the Executive as if performing a provisional classification under sub-paragraph (2).

Special control measures for laboratories, animal rooms and industrial processes

3.– (1) Every employer who is engaged in any of the activities specified in sub-paragraph (3) shall ensure that measures taken to control adequately the exposure of his employees to biological agents include, in particular, the most suitable combination of containment measures from those listed in Parts II and III of this Schedule as appropriate, taking into account –

(a) the nature of the activity specified in sub-paragraph (3);

(b) the minimum containment level specified in sub-paragraph (4);

(c) the risk assessment; and

(d) the nature of the biological agent concerned.

(2) An employer who is engaged in –

(a) any of the activities specified in sub-paragraph (3)(a) or (b) shall select measures from Part II of this Schedule;

(b) the activity specified in sub-paragraph (3)(c) shall select measures from Part III of this Schedule and, subject to sub-paragraph (4), when making that selection he may combine measures from different containment levels on the basis of a risk assessment related to any particular process or part of a process.

(3) The activities referred to in sub-paragraph (1) are –

(a) research, development, teaching or diagnostic work in laboratories which involves working with a Group 2, Group 3 or Group 4 biological agent or material containing such an agent;

(b) working with animals which have been deliberately infected with a Group 2, Group 3 or Group 4 biological agent or which are, or are suspected of being, naturally infected with such an agent; and

(c) industrial processes which involve working with a Group 2, Group 3 or Group 4 biological agent.

(4) Subject to sub-paragraph (5), the minimum containment level referred to in sub-paragraph (1) shall be –

(a) level 2 for activities which involve working with a Group 2 biological agent;

(b) level 3 for activities which involve working with a Group 3 biological agent;

(c) level 4 for activities which involve working with a Group 4 biological agent;

(d) level 2 for laboratories which do not intentionally propagate, concentrate or otherwise increase the risk of exposure to a biological agent but work with materials in respect of which it is unlikely that a Group 3 or Group 4 biological agent is present;

(e) level 3 or 4, where appropriate, for laboratories which do not intentionally propagate, concentrate or otherwise increase the risk of exposure to a Group 3 or Group 4 biological agent but where the employer knows, or it is likely, that such a containment level is necessary; and

(f) level 3 for activities where it has not been possible to carry out a conclusive assessment but where there is concern that the activity might involve a serious health risk for employees.

(5) The Health and Safety Commission may approve guidelines specifying the minimum containment measures which are to apply in any particular case.

(6) The Health and Safety Commission shall not approve any guidelines under paragraph (5) unless it is satisfied that the health of any person who is likely to be affected by the use of those guidelines will not be prejudiced.

List of employees exposed to certain biological agents

4.– (1) Subject to sub-paragraph (2), every employer shall keep a list of employees exposed to a Group 3 or Group 4 biological agent, indicating the type of work done and, where known, the biological agent to which they have been exposed, and records of exposures, accidents and incidents, as appropriate.

(2) Sub-paragraph (1) shall not apply where the results of the risk assessment indicate that –

(a) the activity does not involve a deliberate intention to work with or use that biological agent; and

(b) there is no significant risk to the health of employees associated with that biological agent.

(3) The employer shall ensure that the list or a copy thereof is kept available in a suitable form for at least 40 years from the date of the last entry made in it.

(4) The relevant doctor referred to in regulation 11, and any employee of that employer with specific responsibility for the health and safety of his fellow employees, shall have access to the list.

(5) Each employee shall have access to the information on the list which relates to him personally.

Notification of the use of biological agents

5.– (1) Subject to sub-paragraphs (7) and (8), an employer shall not use for the first time one or more biological agents in Group 2, 3 or 4 at particular premises for any of the activities listed in paragraph 3(3) unless he has –

(a) notified the Executive in writing of his intention to do so at least 20 working days in advance, or such shorter period as the Executive may allow;

(b) furnished with that notification the particulars specified in sub-paragraph (5); and

(c) received the acknowledgement required by sub-paragraph (4).

(2) Subject to sub-paragraphs (7) and (9), an employer shall not use a biological agent which is specified in Part V of this Schedule, except where the use of that agent has been notified to the Executive in accordance with sub-paragraph (1), for any of the activities listed in paragraph 3(3) unless he has –

(a) notified the Executive in writing of his intention to do so at least 20 working days in advance, or such shorter period as the Executive may allow;

(b) furnished with that notification the particulars specified in sub-paragraph (5); and

(c) received the acknowledgement required by sub-paragraph (4).

(3) The Executive may accept a single notification under sub-paragraph (2) in respect of the use of more than one biological agent by the same person.

(4) Upon receipt of the notification required by sub-paragraph (1) or (2), the Executive shall, within 20 working days –

(a) send to the notifier an acknowledgement of receipt; or

(b) if the notification does not contain all of the particulars specified in sub-paragraph (5) –

(i) inform the notifier in writing of the further particulars required, and

(ii) within 10 working days of receipt of those further particulars, send to the notifier an acknowledgement of receipt.

(5) The particulars to be included in the notification referred to in sub-paragraphs (1) and (2) shall be –

(a) the name and address of the employer and the address of the premises where the biological agent will be stored or used;

(b) the name, qualifications and relevant experience of any employee of that employer with specific responsibility for the health and safety of his fellow employees;

(c) the results of the risk assessment;

(d) the identity of the biological agent and, if the agent does not have an approved classification, the Group to which the agent has been assigned; and

(e) the preventive and protective measures that are to be taken.

(6) Where there are changes to processes, procedures or the biological agent which are of importance to health or safety at work and which render the original notification invalid the employer shall notify the Executive forthwith in writing of those changes.

(7) Sub-paragraphs (1) and (2) shall not apply in relation to a biological agent where an intention to use that biological agent has been previously notified to the Executive in accordance with the Genetically Modified Organisms (Contained Use) Regulations 2000[25].

(8) The requirement in sub-paragraph (1) to notify first use of a biological agent in Group 2 or 3 shall not apply to an employer whose only use of that agent is in relation to the provision of a diagnostic service provided that use will not involve a process likely to propagate, concentrate or otherwise increase the risk of exposure to that agent.

(9) The requirement in sub-paragraph (2) to notify use of a biological agent specified in Part V of this Schedule shall not apply to an employer whose only use of that agent is in relation to the provision of a diagnostic service provided that use will not involve a process likely to propagate, concentrate or otherwise increase the risk of exposure to that agent.

Notification of the consignment of biological agents

6.– (1) An employer shall not consign a Group 4 biological agent or anything containing, or suspected of containing, such an agent to any other premises, whether or not those premises are under his ownership or control, unless he has notified the Executive in writing of his intention to do so at least 30 days in advance or before such shorter time as the Executive may approve and with that notification has furnished the particulars specified in sub-paragraph (4).

(2) Sub-paragraph (1) shall not apply where –

(a) the biological agent or material containing or suspected of containing such an agent is being consigned solely for the purpose of diagnosis;

(b) material containing or suspected of containing the biological agent is being consigned solely for the purpose of disposal; or

[25] S.I. 2000/2831.

(c) the biological agent is or is suspected of being present in a human patient or animal which is being transported for the purpose of medical treatment.

(3) Where a Group 4 biological agent is imported into Great Britain, the consignee shall give the notice required by sub-paragraph (1).

(4) The particulars to be included in the notification referred to in sub-paragraph (1) shall be –

(a) the identity of the biological agent and the volume of the consignment;

(b) the name of the consignor;

(c) the address of the premises from which it will be transported;

(d) the name of the consignee;

(e) the address of the premises to which it shall be transported;

(f) the name of the transport operator responsible for the transportation;

(g) the name of any individual who will accompany the consignment;

(h) the method of transportation;

(i) the packaging and any containment precautions which will be taken;

(j) the route which will be taken; and

(k) the proposed date of transportation.

PART II

CONTAINMENT MEASURES FOR HEALTH AND VETERINARY CARE FACILITIES, LABORATORIES AND ANIMAL ROOMS

Containment measures
Containment levels

2
3
4

1 The workplace is to be separated from any other activities in the same building.
No
Yes
Yes

2 Input air and extract air to the workplace are to be filtered using HEPA or equivalent.
No
Yes, on extract air
Yes, on input and double on extract air

3 Access is to be restricted to authorised persons only.
Yes
Yes
Yes, via air-lock key procedure

4 The workplace is to be sealable to permit disinfection.
No
Yes
Yes

5 Specified disinfection procedure.
Yes
Yes
Yes

6 The workplace is to be maintained at an air pressure negative to atmosphere.
No
Yes
Yes

7 Efficient vector control eg rodents and insects.
Yes, for animal containment
Yes, for animal containment
Yes

8 Surfaces impervious to water and easy to clean.
Yes, for bench
Yes, for bench and floor (and walls for animal containment)
Yes, for bench, floor, walls and ceiling

9 Surfaces resistant to acids, alkalis, solvents, disinfectants.
Yes, for bench
Yes, for bench and floor (and walls for animal containment)
Yes, for bench, floor, walls and ceiling

10 Safe storage of biological agents.
Yes
Yes
Yes, secure storage

11 An observation window, or alternative, is to be present, so that occupants can be seen.
No
Yes
Yes

12 A laboratory is to contain its own equipment.
No
Yes, so far as is reasonably practicable
Yes

13 Infected material, including any animal, is to be handled in a safety cabinet or isolator or other suitable containment.
Yes, where aerosol produced
Yes, where aerosol produced
Yes

14 Incinerator for disposal of animal carcases.
Accessible
Accessible
Yes, on site

PART III

CONTAINMENT MEASURES FOR INDUSTRIAL PROCESSES

Containment measures
Containment levels

2
3
4

1 Viable micro-organisms should be contained in a system which physically separates the process from the environment (closed system).
Yes
Yes
Yes

2 Exhaust gases from the closed system should be treated so as to –
Minimise release
Prevent release
Prevent release

3 Sample collection, addition of materials to a closed system and transfer of viable micro-organisms to another closed system, should be performed so as to –
Minimise release
Prevent release
Prevent release

4 Bulk culture fluids should not be removed from the closed system unless the viable micro-organisms have been –
Inactivated by validated means
Inactivated by validated chemical or physical means
Inactivated by validated chemical or physical means

5 Seals should be designed so as to –
Minimise release
Prevent release
Prevent release

6 Closed systems should be located within a controlled area –
Optional
Optional
Yes, and purpose-built

(a) biohazard signs should be posted;
Optional
Yes
Yes

(b) access should be restricted to nominated personnel only;
Optional
Yes
Yes, via air-lock

(c) personnel should wear protective clothing;
Yes, work clothing
Yes
Yes, a complete change

(d) decontamination and washing facilities should be provided for personnel;
Yes
Yes
Yes

(e) personnel should shower before leaving the controlled area;
No
Optional
Yes

(f) effluent from sinks and showers should be collected and inactivated before release;
No
Optional
Yes

(g) the controlled area should be adequately ventilated to minimise air contamination;
Optional
Optional
Yes

(h) the controlled area should be maintained at an air pressure negative to atmosphere;
No
Optional
Yes

(i) input and extract air to the controlled area should be HEPA filtered;
No
Optional
Yes

(j) the controlled area should be designed to contain spillage of the entire contents of closed system;
Optional
Yes
Yes

(k) the controlled area should be sealable to permit fumigation.
No
Optional
Yes

7 Effluent treatment before final discharge.
Inactivated by validated means
Inactivated by validated chemical or physical means
Inactivated by validated physical means

PART IV

BIOHAZARD SIGN

The biohazard sign required by regulation 7(6)(a) shall be in the form shown below –

PART V

BIOLOGICAL AGENTS WHOSE USE IS TO BE NOTIFIED IN ACCORDANCE WITH PARAGRAPH 5(2) OF PART I OF THIS SCHEDULE

Any Group 3 or 4 agent.

The following Group 2 agents:

Bordetella pertussis
Corynebacterium diphtheriae
Neisseria meningitidis

SCHEDULE 4

Regulation 9(2)(a)

FREQUENCY OF THOROUGH EXAMINATION AND TEST OF LOCAL EXHAUST VENTILATION PLANT USED IN CERTAIN PROCESSES

Column 1
Column 2

Process
Minimum frequency

Processes in which blasting is carried out in or incidental to the cleaning of metal castings, in connection with their manufacture.
1 month

Processes, other than wet processes, in which metal articles (other than of gold, platinum or iridium) are ground, abraded or polished using mechanical power, in any room for more than 12 hours in any week.
6 months

Processes giving off dust or fume in which non-ferrous metal castings are produced.
6 months

Jute cloth manufacture.
1 month

SCHEDULE 5

Regulation 10(4)

SPECIFIC SUBSTANCES AND PROCESSES FOR WHICH MONITORING IS REQUIRED

Column 1
Column 2

Substance or process
Minimum frequency

Vinyl chloride monomer.
Continuous or in accordance with a procedure approved by the Health and Safety Commission.

Spray given off from vessels at which an electrolytic chromium process is carried on, except trivalent chromium.
Every 14 days while the process is being carried on.

SCHEDULE 6

Regulation 11(2)(a) and (5)

MEDICAL SURVEILLANCE

Column 1
Column 2

Substances for which medical surveillance is appropriate
Process

Vinyl chloride monomer (VCM).
In manufacture, production, reclamation, storage, discharge, transport, use or polymerisation.

Nitro or amino derivatives of phenol and of benzene or its homologues.
In the manufacture of nitro or amino derivatives of phenol and of benzene or its homologues and the making of explosives with the use of any of these substances.

Potassium or sodium chromate or dichromate.
In manufacture.

Ortho-tolidine and its salts. Dianisidine and its salts. Dichlorobenzidine and its salts.
In manufacture, formation or use of these substances.

Auramine. Magenta.
In manufacture.

Carbon disulphide. Disulphur dichloride. Benzene, including benzol. Carbon tetrachloride. Trichlorethylene.
Processes in which these substances are used, or given off as vapour, in the manufacture of indiarubber or of articles or goods made wholly or partially of indiarubber.

Pitch.
In manufacture of blocks of fuel consisting of coal, coal dust, coke or slurry with pitch as a binding substance.

SCHEDULE 7

Regulation 12(5)

LEGISLATION CONCERNED WITH THE LABELLING OF CONTAINERS AND PIPES

The Chemicals (Hazard Information and Packaging for Supply) Regulations 2002 (CHIP) (SI 2002/1689);

The Health and Safety (Safety Signs and Signals) Regulations 1996 (SI 1996/341);

The Radioactive Material (Road Transport) Regulations 2002 (SI 2002/1093);

The Carriage of Dangerous Goods by Rail Regulations 1996 (SI 1996/2089);

The Packaging, Labelling and Carriage of Radioactive Material by Rail Regulations 2002 (SI 2002/2099);

The Carriage of Dangerous Goods (Classification, Packaging and Labelling) and Use of Transportable Pressure Receptacles Regulations 1996 (SI 1996/2092);

The Carriage of Explosives by Road Regulations 1996 (SI 1996/2093);

The Carriage of Dangerous Goods by Road Regulations 1996 (SI 1996/2095); and

The Good Laboratory Practice Regulations 1999 (SI 1999/3106).

SCHEDULE 8

Regulation 14(1)

FUMIGATIONS EXCEPTED FROM REGULATION 14

Column 1
Column 2

Fumigant
Nature of fumigation

Hydrogen cyanide.
Fumigations carried out for research.

Fumigations in fumigation chambers.

Fumigations in the open air to control or kill mammal pests.

Methyl bromide.

Fumigations carried out for research.

Fumigations in fumigation chambers.

Fumigations of soil outdoors under gas-proof sheeting where not more than 1000 kg is used in any period of 24 hours on the premises.

Fumigations of soil under gas-proof sheeting in glasshouses where not more than 500 kg is used in any period of 24 hours on the premises.

Fumigations of compost outdoors under gas-proof sheeting where not more than 10 kg of methyl bromide is used in any period of 24 hours on the premises.

Fumigations under gas-proof sheeting inside structures other than glasshouses and mushroom houses where not more than 5 kg of methyl bromide is used in each structure during any period of 24 hours.

Fumigations of soil or compost in mushroom houses where not more than 5 kg of methyl bromide is used in any one fumigation in any period of 24 hours.

Fumigations of containers where not more than 5 kg of methyl bromide is used in any one fumigation in a period of 24 hours.

Phosphine.
Fumigations carried out for research.

Fumigations in fumigation chambers.

Fumigations under gas-proof sheeting inside structures where not more than 1 kg of phosphine in each structure is used in any period of 24 hours.

Fumigations in containers where not more than 0.5 kg of phosphine is used in any one fumigation in any period of 24 hours.

Fumigations in individual impermeable packages.

Fumigations in the open air to control or kill mammal pests.

SCHEDULE 9

Regulation 14(2)

NOTIFICATION OF CERTAIN FUMIGATIONS

PART I

PERSONS TO WHOM NOTIFICATIONS MUST BE MADE

1. In the case of a fumigation to be carried out within the area of a harbour authority, advance notification of fumigation shall, for the purposes of regulation 14(2)(a), be given to –

   (a) that authority;

   (b) an inspector appointed under section 19 of the 1974 Act, if that inspector so requires; and

   (c) where the fumigation –

      (i) is to be carried out on a sea-going ship, the chief fire officer of the area in which the ship is situated and the officer in charge of the office of Her Majesty's Customs and Excise at the harbour, or

      (ii) is the space fumigation of a building, the chief fire officer of the area in which the building is situated.

2. In the case of a fumigation, other than a fumigation to which paragraph (1) applies, advance notification of fumigation shall be given to –

   (a) the police officer for the time being in charge of the police station for the police district in which the fumigation is carried out;

   (b) an inspector appointed under section 19 of the 1974 Act, if that inspector so requires; and

   (c) where the fumigation is to be carred out on a sea-going ship or is the space fumigation of a building, the chief fire officer of the area in which the ship or building is situated.

PART II

INFORMATION TO BE GIVEN IN ADVANCE NOTICE OF FUMIGATIONS

3. The information to be given in a notification made for the purposes of regulation 14(2) shall include the following –

   (a) the name, address and place of business of the fumigator and his telephone number;

   (b) the name of the person requiring the fumigation to be carried out;

   (c) the address and description of the premises where the fumigation is to be carried out;

   (d) the date on which the fumigation is to be carried out and the estimated time of commencement and completion;

   (e) the name of the operator in charge of the fumigation; and

   (f) the fumigant to be used.

EXPLANATORY NOTE

(This note is not part of the Regulations)

1. These Regulations re-enact, with modifications, the Control of Substances Hazardous to Health Regulations 1999 (S.I. 1999/437) ("the 1999 Regulations"). The 1999 Regulations imposed duties on employers to protect employees and other persons who may be exposed to substances hazardous to health and also imposed certain duties on employees concerning their own protection from such exposure, and prohibited the import into the United Kingdom of certain substances and articles from outside the European Economic Area.

2. The Regulations, with the exception of regulations 8, 9, 11(8), (10) and (11) and 14, implement as respects Great Britain –

   (a) Council Directive 78/610/EEC (OJ No. L 197, 22.7.78, p.12) on the approximation of the laws, regulations and administrative provisions of the Member States on the protection of the health of workers exposed to vinyl chloride monomer,

   (b) point 3 of Article 1 of Council Directive 89/677/EEC (OJ No. L 398, 30.12.89, p.19) amending for the 8th time the Marketing and Use Directive insofar as that point relates to the importation, supply or use of benzene and substances containing benzene for such purposes,

(c) Council Directive 90/394/EEC (OJ No. L 196, 26.7.90, p.1) on the protection of workers from risks related to exposure to carcinogens at work (sixth individual Directive within the meaning of Article 16(1) of Directive 89/391/EEC) insofar as it relates to carcinogens other than asbestos,

(d) that part of Commission Directive 96/55/EC (OJ No. L 231, 12.9.96, p.20) adapting to technical progress for the 2nd time Annex I to Council Directive 76/769/EEC ("the Marketing and Use Directive") (OJ No. L 262, 27.9.76, p.201) on the approximation of the laws, regulations and administrative provisions of the Member States relating to restrictions on the marketing and use of certain dangerous substances and preparations which relates to supply of specified substances for use at work,

(e) Council Directive 98/24/EC (OJ No. L 131, 5.5.98, p.11) on the protection of the health and safety of workers from risks related to chemical agents at work (fourteenth individual Directive within the meaning of Article 16(1) of Directive 89/391/EEC) insofar as it relates to risks to health from exposure to substances other than asbestos or lead, and

(f) Directive 2000/54/EC (OJ No. L 262, 17.10.2000, p.21) of the European Parliament and of the Council on the protection of workers from risks related to exposure to biological agents at work (seventh individual Directive within the meaning of Article 16(1) of Directive 89/391/EEC).

3. The Regulations are consistent with the provisions of Commission Directive 91/322/EEC (OJ No. L 177, 5.7.91, p.22) on establishing indicative limit values by implementing Council Directive 80/1107/EEC on the protection of workers from risks related to exposure to chemical, physical and biological agents at work.

4. In addition to minor and drafting amendments, the Regulations make the following changes of substance –

(a) include certain further definitions (regulation 2);

(b) extend the matters to be considered when carrying out an assessment of the risk from exposure to substances hazardous to health (regulation 6);

(c) detail the measures which the employer must take to prevent or adequately control the exposure of his employees to substances hazardous to health (regulation 7);

(d) provide for further duties in respect of care and decontamination of personal protective equipment (regulation 9);

(e) provide for the keeping of an individual record of air monitoring where an employee is required to be under health surveillance (regulation 10);

(f) extend the duties on employers with respect to health surveillance where an employee is found to have an identifiable disease or adverse health effect caused by exposure to a substance hazardous to health (regulation 11);

(g) introduce a duty to ensure that the contents of containers and pipes for substances hazardous to health used at work are clearly identifiable (regulation 12(5));

(h) introduce a duty on the employer to prepare procedures, provide information and establish warning systems to deal with an emergency in the workplace related to the presence of a substance hazardous to health (regulation 13); and

(i) apply the extension to the meaning of "work" in Part I of the 1974 Act to all Group 2, 3, or 4 biological agents and transfer certain provisions relating to biological agents which had been in Schedule 3 of the 1999 Regulations to regulations 7 and 12.

5. Copies of the publications mentioned in the Regulations are obtainable as follows –

(a) a list of the maximum exposure limits and occupational exposure standards which the Health and Safety Commission has approved is available in the publication "EH40, Occupational Exposure Limits" obtainable from HSE Books, PO Box 1999, Sudbury, Suffolk CO10 2WA; and

(b) British Standard BS EN 481 1993, referred to in regulation 2(1), relating to workplace atmospheres is obtainable from the British Standards Institution, BSI House, 389 Chiswick High Road, London W4 4AL.

6. A copy of the regulatory impact assessment prepared in respect of these Regulations can be obtained from the Health and Safety Executive, Economic Advisers Unit, Rose Court, 2 Southwark Bridge, London SE1 9HS. A copy of the transposition note in relation to implementation of the Directives set out in paragraph 2 can be obtained from the Health and Safety Executive, International Branch at the same address. Copies of both these documents have been placed in the Library of each House of Parliament.

INDEX

- Digests all other key sources
- Organised by injury
- Multiplier applied including automatic *Heil* v *Rankin* calculator (also usable independently on your caseload)
- Portable one volume paperback
- Updated monthly by download, ensuring you are using the latest version even on a laptop

Book and 2003 Service £149.00

Network Licence £10 + VAT per user

ISBN 1 85811 264 8

Paperback and free single user licence to electronic service

kevan on credit hire

Tim Kevan, Barrister, 1 Temple Gardens

In road traffic claims, insurance companies compensate individuals in the form of repair and/or a replacement car. A 'credit hire' and 'credit repair' industry has developed on the basis that the insurer will pay. Whether you are a claimant or defendant practitioner, this book will clearly and succinctly ensure that keep up to date with the law – including Consumer Credit issues.

£48 2002 ISBN 1 85811 275 3

children and personal injury

Andrew Hogan, Barrister, Ropewalk Chambers

This succinct and well-written book looks at the law and procedure governing the litigation of claims brought by injured children, covering the unique aspects of acting for children.

£28 2000 1 85811 262 1

medical records for lawyers

Ali Malsher BA (Hons), MA, Solicitor, Registered General Nurse

Medical areas are covered with basic anatomy and physiology, common complaints, symptoms, investigations and tests and explanation of treatments and sample medical records. You will also be able to cross-reference records to check that certain activities have been completed by the medical and nursing teams. Medical terminology, medical words and symbols are all explained succinctly.

£42 2002 ISBN 1 85811 259 1

Printed in the United Kingdom
by Lightning Source UK Ltd.
104991UKS00002B/6